Applied Probability
Control
Economics
Information and Communication
Modeling and Identification
Numerical Techniques
Optimization

Applications of
Mathematics

20

Edited by   A. V. Balakrishnan

Applications of Mathematics

1 Fleming/Rishel, **Deterministic and Stochastic Optimal Control** (1975)
2 Marchuk, **Methods of Numerical Mathematics,** Second Ed. (1982)
3 Balakrishnan, **Applied Functional Analysis,** Second Ed. (1981)
4 Borovkov, **Stochastic Processes in Queueing Theory** (1976)
5 Lipster/Shiryayev, **Statistics of Random Processes I: General Theory** (1977)
6 Lipster/Shiryayev, **Statistics of Random Processes II: Applications** (1978)
7 Vorob'ev, **Game Theory: Lectures for Economists and Systems Scientists** (1977)
8 Shiryayev, **Optimal Stopping Rules** (1978)
9 Ibragimov/Rozanov, **Gaussian Random Processes** (1978)
10 Wonham, **Linear Multivariable Control: A Geometric Approach,** Third Ed. (1985)
11 Hida, **Brownian Motion** (1980)
12 Hestenes, **Conjugate Direction Methods in Optimization** (1980)
13 Kallianpur, **Stochastic Filtering Theory** (1980)
14 Krylov, **Controlled Diffusion Processes** (1980)
15 Prabhu, **Stochastic Storage Processes: Queues, Insurance Risk, and Dams** (1980)
16 Ibragimov/Has'minskii, **Statistical Estimation: Asymptotic Theory** (1981)
17 Cesari, **Optimization: Theory and Applications** (1982)
18 Elliott, **Stochastic Calculus and Applications** (1982)
19 Marchuk/Shaidourov, **Difference Methods and Their Extrapolations** (1983)
20 Hijab, **Stabilization of Control Systems** (1986)

O. Hijab

Stabilization of Control Systems

Springer-Verlag
New York Berlin Heidelberg
London Paris Tokyo

O. Hijab
Mathematics Department
Temple University
Philadelphia, PA 19122
U.S.A.

Managing Editor
A. V. Balakrishnan
Systems Science Department
University of California
Los Angeles, CA 90024
U.S.A.

AMS Classification: 93EXX

With 3 Illustrations

Library of Congress Cataloging in Publication Data
Hijab, O.
  Stabilization of control systems.
  (Applications of mathematics; 20)
  Includes bibliographical references and index.
  1. System analysis. 2. Stochastic systems.
3. Stability. I. Title. II. Series.
QA402.H55 1986   003   86-13920

Typeset by Asco Trade Typesetting Ltd., Hong Kong.
Printed and bound by R. R. Donnelley & Sons, Harrisonburg, Virginia.
Printed in the United States of America.

9 8 7 6 5 4 3 2 1

ISBN 0-387-96384-7 Springer-Verlag New York Berlin Heidelberg
ISBN 3-540-96384-7 Springer-Verlag Berlin Heidelberg New York

To Carol Armstrong,
who makes it all worthwhile

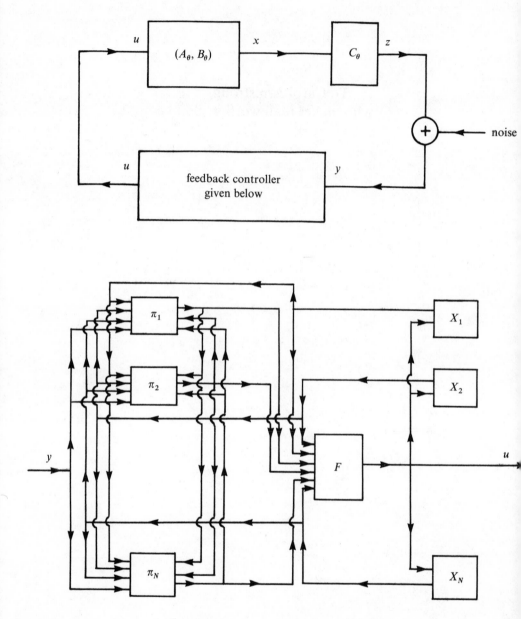

Feedback Control Appearing in Chapter 5

Contents

Introduction   ix

Notation   xi

CHAPTER 1
Input/Output Properties   1

1.1. An Example   1
1.2. Review of Linear Algebra   4
1.3. Linear Systems   8
1.4. Controllability and Observability   10
1.5. Minimality   13
1.6. Realizability   16
1.7. Notes and References   19

CHAPTER 2
The LQ Regulator   21

2.1. Stabilization   21
2.2. Properness   24
2.3. Optimal Control   28
2.4. The Riccati Equation   32
2.5. The Space $M(m, n, p)$ 35
2.6. Notes and References   41

CHAPTER 3
Brownian Motion   43

3.1. Preliminary Definitions   43
3.2. Stochastic Calculus   52
3.3. Cameron–Martin–Girsanov Formula   57
3.4. Notes and References   62

CHAPTER 4
Filtering   64

4.1. Filtering   64
4.2. Consistency   73
4.3. Shannon Information   77
4.4. Notes and References   82

CHAPTER 5
The Adaptive *LQ* Regulator   84

5.1. Introduction   84
5.2. Smooth Admissible Controls   87
5.3. Adaptive Stabilization   91
5.4. Optimal Control   94
5.5. Bellman Equation   98
5.6. Notes and References   102

APPENDIX
Solutions to Exercises   103

Index   127

Introduction

The problem of controlling or stabilizing a system of differential equations in the presence of random disturbances is intuitively appealing and has been a motivating force behind a wide variety of results grouped loosely together under the heading of "Stochastic Control."

This book is concerned with a special instance of this general problem, the "Adaptive LQ Regulator," which is a stochastic control problem of partially observed type that can, in certain cases, be solved explicitly. We first describe this problem, as it is the focal point for the entire book, and then describe the contents of the book.

The problem revolves around an uncertain linear system

$$\dot{x} = A_\theta x + B_\theta u, \qquad x(0) = x_\theta^0 \text{ in } \mathbb{R}^n,$$

where $\theta \in \{1, \ldots, N\}$ is a random variable representing this uncertainty and (A_j, B_j, C_j) and x_j^0 are the coefficient matrices and initial state, respectively, of a linear control system, for each $j = 1, \ldots, N$. A common assumption is that the mechanism causing this uncertainty is additive noise, and that consequently the "controller" has access only to the observation process $y(\cdot)$ where

$$y = C_\theta x + \dot{\eta}.$$

Here $\dot{\eta}(\cdot)$ is the proverbial "white noise," the fictional (time) derivative of Brownian motion. The problem then is to seek a causal feedback

$$u(t) = F(t, y(s), 0 \le s \le t), \qquad t \ge 0,$$

that minimizes a cost functional of the type

$$J(u) = E\left(\int_0^\infty |u(t)|^2 + |x(t)|^2 \, dt\right) + H(\infty)$$

over all (Borel) causal feedbacks F. An immediate consequence is then that any $u(\cdot)$ whose corresponding cost $J(u)$ is finite is necessarily stabilizing

$$P(x(t) \to 0 \text{ as } t \uparrow \infty) = 1.$$

Here $H(\infty)$ is a bounded perturbation that is a measure of the amount of information concerning θ available in the process $y(\cdot)$. *The purpose of this book is to present an exposition of the mathematical theory necessary to deal with the above problem.*

The book falls naturally into two parts: the first is Chapters 1 and 2, and is a quick course on basic linear system theory which is (almost by definition) the material necessary to solve the above problem in the special case $N = 1$ (there is no probability theory in this case). Although this material has been well known for some time, a presentation of these results that moves in a clear and concise fashion is apparently unavailable. An exception is Brockett's 1970 text [1.1], which is at present out of print.

The second part of the book, Chapters 3, 4, and 5, is of a more advanced nature. The presentation here is quicker. By contrast, Chapters 1 and 2 are developed at a leisurely pace. We do this because we feel that otherwise the student "does not see the forest for the trees." Nevertheless, we have attempted to include all relevant material, some of it stated but not proved (Sections 3.1 and 3.2). This method allows material to be expanded, by consulting the references, or deleted, at the reader's discretion. Chapter 3 reviews the basic probability theory needed and in particular the construction of the stochastic integral. Here we have systematically followed the exposition of Stroock and Varadhan [3.8]. Chapter 4 deals with the relevant class of filtering problems, and Chapter 5 combines the results of Chapters 2 and 4 in an analysis of the Adaptive LQ Regulator problem. One by-product of following the presentation of [3.8] is that the formulations of the filtering results in Chapter 4 are somewhat sharper than those which usually appear in the literature.

This book is meant for use as a text. Because of this, we have included over 120 exercises. Apart from their tutorial value, their presence streamlines many of the proofs. Thus they are an integral part of the text. An appendix is included in which solutions to the exercises are provided. The review sections 1.2 and 3.1 are almost all exercises. The bibliographical references are grouped into five sections, one at the end of each chapter. Thus reference [3.8] denotes the eighth reference at the end of Chapter 3.

I want to thank the colleagues and friends who have supported and encouraged me over the years. R. W. Brockett, W. H. Fleming, M. Hazewinkel, R. Hermann, A. J. Krener, and G. C. Papanicolaou stimulated and encouraged me when I started out. To them I am deeply indebted. The support of the National Science Foundation (NSF/DMS/8418885) is gratefully acknowledged. Special thanks go to L. E. Clemens who systematically went through the manuscript and helped me overcome many obstacles during the writing of this book.

Notation

		Section
A, B, C	system matrices	1.2
b, c, x, y, u	vectors	1.2
$|c|, |x|, |A|$	Euclidean norms	1.2
L	solutions of the Lyapunov equation	1.2
$u(\cdot), v(\cdot)$	controls	1.3
$x^u(\cdot)$	state trajectory	1.3
$y^u(\cdot)$	output trajectory	1.3
$u *_T v$	concatenation of $u(\cdot)$ and $v(\cdot)$ at time T	1.3
$G(s)$	a transfer function	1.3
$g(s)$	a scalar transfer function	1.3
$I/O(x)$	input–output map starting from x	1.3
V^T	controllable subspace	1.4
W^T	unobservable subspace	1.4
F	feedback matrix	2.1
$\bar{A}$	stable matrix	2.1
$J^u(x)$	cost associated to $u(\cdot)$ starting from x	2.1
$S(x)$	minimum cost starting from x	2.2
K	solution of the algebraic Riccati equation	2.3
$\bar{G}(s), G^\#(s), \bar{G}^\#(s)$	associated transfer functions	2.3
$\tilde{G}(s)$	Hamiltonian transfer fuction	2.3
$M(m, n, p)$	space of p by m transfer functions of dimension n	2.5
Ω	set of elementary events	3.1
$\mathscr{F}, \mathscr{B}, \mathscr{M}, \mathscr{X}, \mathscr{Y}, \mathscr{D}$	σ-algebras	3.1
τ, σ	stopping times	3.1
P, Q	probability measures	3.1

$a \wedge b$	$\min(a, b)$	3.1
E^P	expectation against P	3.1
R	Radon–Nikodym derivative	3.1
$\ll$	absolutely continuous	3.1
$g_m(t, x)$	Gauss–Weierstrass kernel	3.1
$\eta(\cdot)$	a Brownian motion	3.1
$R(\cdot)$	a martingale	3.1
π	a probability distribution on $\{1, \ldots, N\}$	3.1
$C([0, \infty); \mathbb{R}^m)$	path space	3.1
W	Wiener measure	3.1
$\log^+ a$	$\max(\log a, 0)$	3.1
L	diffusion generator	3.2
$[\ ]$	greatest integer function	3.2
f, g	vector fields	3.2
ζ	explosion time	3.2
$z(\cdot)$	a signal process	4.1
$y(\cdot)$	an observation process	4.1
θ	state parameter	4.1
$l_j(\cdot)$	likelihood functional	4.1
$\pi_j(\cdot)$	conditional probability	4.1
$\hat{z}(\cdot)$	conditional expectation of the signal	4.1
$v(\cdot)$	innovations process	4.1
$\hat{\varphi}$	Fourier transform of φ	4.1
$I(\mu; v)$	information functional	4.3
$I(\pi; \pi^0)$	information functional on $\{1, \ldots, N\}$	4.3
$I(x_1, x_2)$	Shannon information of the pair x_1, x_2	4.3
$I(T)$	Shannon information of the pair θ and $y(t), 0 \leq t \leq T$	4.3
$u(\cdot)$	a control process	5.1
$x^u(\cdot)$	state trajectory process	5.1
$y^u(\cdot)$	observation process	5.1
$J^u(x, \pi)$	cost starting from x, π	5.1
$I^u(\infty)$	Shannon information of θ and $y^u(\cdot)$	5.1
F, f	feedback functions	5.2
$S(x, \pi)$	minimum cost starting from x, π	5.4
G^u	diffusion generator of $(x^u(\cdot), \pi^u(\cdot))$	5.5
$\mathscr{P}$	probability simplex	5.5
H	entropy function	5.5

CHAPTER 1
Input/Output Properties

1.1. An Example

Consider an idealization of a point mass in the presence of an inverse square force field $-k/r^2$.

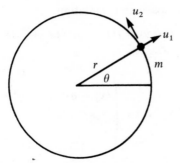

The mass m is equipped with the ability to exert a thrust u_1 in the radial direction and a thrust u_2 in the tangential direction. We derive the equations of motion of this system.

The Lagrangian here is $L = T - V$, where

$$T = \text{kinetic energy} = \tfrac{1}{2}m(\dot{r}^2 + r^2\dot{\theta}^2),$$

$$V = \text{potential energy} = -\frac{k}{r}.$$

Lagrange's equations in the coordinate q read

$$\frac{d}{dt}\left(\frac{\partial L}{\partial \dot{q}}\right) - \frac{\partial L}{\partial q} = \text{external force in the } q \text{ direction.}$$

This leads to

$$m\ddot{r} - mr\dot{\theta}^2 + \frac{k}{r^2} = u_1,$$

$$2r\dot{r}\dot{\theta}m + r^2\ddot{\theta}m = ru_2.$$

One solution of these equations, when no thrust is exerted ($u_1 = u_2 = 0$), is

$$r(t) = \sigma, \qquad \theta(t) = \omega t \qquad \text{with} \qquad \frac{k}{m} = \sigma^3\omega^2.$$

To study the behavior of the system near this circular orbit, change coordinates to

$$x_1 = r - \sigma, \qquad x_2 = \dot{r}, \qquad x_3 = \theta - \omega t, \qquad x_4 = \dot{\theta} - \omega.$$

This leads to

$$\dot{x}_1 = x_2,$$

$$\dot{x}_2 = (x_1 + \sigma)(x_4 + \omega)^2 - \frac{k}{m(x_1 + \sigma)^2} + \frac{u_1}{m},$$

$$\dot{x}_3 = x_4,$$

$$\dot{x}_4 = -\frac{2x_2(x_4 + \omega)}{x_1 + \sigma} + \frac{u_2}{m(x_1 + \sigma)},$$

which leads to the state-space representation

$$\dot{x} = f(x, u) = f(x) + g_1(x)u_1 + g_2(x)u_2$$

with

$$f(x) = f(x_1, x_2, x_3, x_4) = \begin{pmatrix} x_2 \\ (x_1 + \sigma)(x_4 + \omega)^2 - \dfrac{k}{m(x_1 + \sigma)^2} \\ x_4 \\ -\dfrac{2x_2(x_4 + \omega)}{x_1 + \sigma} \end{pmatrix},$$

$$g_1(x) = g_1(x_1, x_2, x_3, x_4) = \begin{pmatrix} 0 \\ \dfrac{1}{m} \\ 0 \\ 0 \end{pmatrix},$$

$$g_2(x) = g_2(x_1, x_2, x_3, x_4) = \begin{pmatrix} 0 \\ 0 \\ 0 \\ 1 \\ \dfrac{}{m(x_1 + \sigma)} \end{pmatrix}.$$

Note that f, g_1, and g_2 are defined on open subsets of $\mathbb{R}^4$ that include the origin. If we are only interested in the position of the mass, then we set the outputs of the system to be

$$y_1 = r - \sigma \quad \text{and} \quad y_2 = \theta - \omega t,$$

which can be written in state-space form

$$y = h(x),$$

with

$$h(x) = h(x_1, x_2, x_3, x_4) = \begin{pmatrix} h_1(x) \\ h_2(x) \end{pmatrix} = \begin{pmatrix} x_1 \\ x_3 \end{pmatrix}.$$

Let

$$x_\varepsilon^0 \sim x_0^0 + \varepsilon x_1^0 + \cdots,$$

$$u_\varepsilon(t) \sim u_0(t) + \varepsilon u_1(t) + \cdots$$

be a perturbation of a given control $u_0(\cdot)$ and initial state x_0^0. The *linearization* of the system

$$\dot{x} = f(x, u), \qquad x(0) = x^0 \text{ in } \mathbb{R}^n, \tag{1.1}$$

$$y = h(x), \tag{1.2}$$

near the trajectory corresponding to the given inputs $(x_0^0, u_0(\cdot))$ is as follows. Let $x_\varepsilon(t)$, $0 \le t \le T$, be the solution of (1.1), let $y_\varepsilon(t)$, $0 \le t \le T$, be the output of (1.2), corresponding to $x^0 = x_\varepsilon^0$ and $u = u_\varepsilon$. Expand $x_\varepsilon(t)$ in powers of ε, as well as $y_\varepsilon(t)$,

$$x_\varepsilon(t) \sim x_0(t) + \varepsilon x_1(t) + \cdots,$$

$$y_\varepsilon(t) \sim y_0(t) + \varepsilon y_1(t) + \cdots.$$

Inserting this into (1.1), expanding $f(x, u)$ and $h(x)$ in a Taylor series about $(x_0(t), u_0(t))$, and equating like powers of ε yields

$$\dot{x}_1 = A(t)x_1 + B(t)u_1, \qquad x_1(0) = x_1^0, \tag{1.3}$$

$$y_1 = C(t)x_1, \tag{1.4}$$

where

$$A(t) = (a_{ij}(t)), \qquad a_{ij}(t) = \frac{\partial f_i}{\partial x_j}(x_0(t), u_0(t)), \qquad i, j = 1, \ldots, n,$$

$$B(t) = (b_{ij}(t)), \qquad b_{ij}(t) = \frac{\partial f_i}{\partial u_j}(x_0(t), u_0(t)), \qquad i = 1, \ldots, n, \quad j = 1, \ldots, m,$$

$$C(t) = (c_{ij}(t)), \qquad c_{ij}(t) = \frac{\partial h_i}{\partial x_j}(x_0(t)), \qquad i = 1, \ldots, p, \quad j = 1, \ldots, n.$$

The system (1.3), (1.4) is the linearization of the system (1.1), (1.2). In particular, if

$$f(x, u) = f(x) + g_1(x)u_1 + \cdots + g_m(x)u_m$$

and $u_0(t) = 0, 0 \leq t \leq T, f(x^0) = 0, h(x^0) = 0$, then

$$A = (a_{ij}) = \left(\frac{\partial f_i}{\partial x_j}(x^0)\right), \qquad B = (b_{ij}) = (g_i(x^0)_j), \qquad C = (c_{ij}) = \left(\frac{\partial h_i}{\partial x_j}(x^0)\right).$$

For the above example $n = 4, m = 2, p = 2$ and

$$A = \begin{pmatrix} 0 & 1 & 0 & 0 \\ 3\omega^2 & 0 & 0 & 2\sigma\omega \\ 0 & 0 & 0 & 1 \\ 0 & -2\dfrac{\omega}{\sigma} & 0 & 0 \end{pmatrix}, \qquad B = (b_1, b_2) = \begin{pmatrix} 0 & 0 \\ \dfrac{1}{m} & 0 \\ 0 & 0 \\ 0 & \dfrac{1}{m\sigma} \end{pmatrix},$$

$$C = \begin{pmatrix} 1 & 0 & 0 & 0 \\ 0 & 0 & 1 & 0 \end{pmatrix}.$$

1.2. Review of Linear Algebra

Throughout we will adopt the following notation. The set of complex (real) numbers will be denoted by $\mathbb{C}(\mathbb{R})$. x, u, and y denote column vectors in $\mathbb{C}^n$, $\mathbb{C}^m$, and $\mathbb{C}^p$, respectively, while A, B, and C denote n by n, n by m, and p by n matrices with complex entries. In the case $m = 1$ or $p = 1$, B or C will be denoted by the corresponding lowercase letters b or c. Note that c is then a row vector. An asterisk $*$ denotes the adjoint (conjugate transpose) of a matrix. The conjugate of the complex number z is z^*. Thus $A = (a_{ij})$ implies $A^* = (a_{ji}^*)$. The norm squared of the vector x is $|x|^2 = x^*x$, of the row vector c is $|c|^2 = cc^*$, and of the matrix A is $|A|^2 = \text{trace}(A^*A)$. In all three cases this is the sum of the squares of the absolute values of the entries of x, c, and A. Unless otherwise specified, vectors will always be column vectors. A matrix A is self-adjoint if $A = A^*$. Note that for any n by n matrices A and A' and any n by 1 vector x, $|AA'| \leq |A||A'|$ and $|Ax| \leq |A||x|$.

The *eigenpolynomial* corresponding to an n by n matrix A is the (monic, degree n) polynomial $\det(sI - A)$. The roots of this polynomial are the eigenvalues of A, $\lambda_1, \ldots, \lambda_r, r \leq n$,

$$\det(sI - A) = \prod_{j=1}^{r} (s - \lambda_j)^{m_j};$$

here m_j is the multiplicity of λ_j. An eigenvector corresponding to the eigenvalue λ is a nonzero vector x in $\mathbb{C}^n$ satisfying $Ax = \lambda x$.

By Cramer's rule, the *resolvent*

$$(sI - A)^{-1} = \frac{1}{q(s)}P(s),$$

where $q(s)$ is the eigenpolynomial of A and $P(s)$ is a matrix whose entries are polynomials in s of degree at most $n - 1$.

Given a matrix $G(s)$ of rational functions of a complex variable s, a pole of $G(s)$ is a complex number for which the denominator of some entry of $G(s)$ vanishes. For example, the eigenvalues of A are precisely the poles of $G(s) = (sI - A)^{-1}$.

Let C denote a closed contour in the complex plane encircling all the eigenvalues of A.

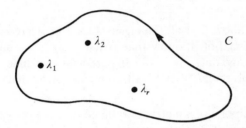

Let $f(s)$ be an entire function. The matrix $f(A)$ is defined by

$$f(A) = \frac{1}{2\pi i}\int_C f(s)(sI - A)^{-1}\,ds, \qquad i = \sqrt{-1}. \tag{2.1}$$

Applying Cauchy's theorem to each entry of the matrix appearing inside the integral, we see that $f(A)$ does not depend on the choice of the contour C. In particular, if C is chosen to be a circle so large that, for s on C, $|A|/|s|$ is less than one, then

$$(sI - A)^{-1} = \frac{1}{s}I + \frac{1}{s^2}A + \frac{1}{s^3}A^2 + \cdots \tag{2.2}$$

is uniformly convergent and so multiplying by $f(s)$ and integrating over C,

$$f(A) = a_0 I + a_1 A + a_2 A^2 + \cdots, \tag{2.3}$$

where $a_k = (1/2\pi i)\int_C f(s)s^{-k-1}\,ds$, $k = 0, 1, 2, \ldots$. Appealing to Cauchy's formula, we have $k!\,a_k = f^{(k)}(0)$. Thus

$$f(s) = a_0 + a_1 s + a_2 s^2 + \cdots$$

which shows that $f(A)$ is obtained from $f(s)$ by "plugging in" A.

The residue theorem now implies

$$f(A) = \sum_{j=1}^{r} \frac{1}{(m_j - 1)!}\left(\frac{d}{ds}\right)^{m_j - 1}\left(f(s)(sI - A)^{-1}(s - \lambda_j)^{m_j}\right)_{s=\lambda_j}.$$

1.2.1. Exercise. Show that for any entire functions $f(s)$, $g(s)$ one has

$$f(A)g(A) = (fg)(A) = g(A)f(A),$$
$$f(A) + g(A) = (f + g)(A).$$

Equation (2.1) also implies the *Cayley–Hamilton* theorem: if $q(s)$ is the eigen-polynomial of A then $q(A) = 0$. This follows from (2.1) since $q(s)(sI - A)^{-1}$ is a matrix of entire functions and hence by Cauchy's theorem the integral in (2.1) vanishes.

More generally, the *minimal polynomial* of A is the monic polynomial of least degree $m(s)$ satisfying $m(A) = 0$. Given a matrix $G(s)$ of rational functions of s, the *denominator* of $G(s)$ is the monic polynomial $m(s)$ of least degree such that $m(s)G(s)$ is a matrix with polynomial entries.

1.2.2. Exercise. Show that all the eigenvalues λ of A are roots of $m(s)$, the minimal polynomial of A. Show that $p(s)(sI - A)^{-1}$ is a matrix with polynomial entries if and only if $p(A) = 0$. In particular, conclude that the denominator of the resolvent $(sI - A)^{-1}$ is the minimal polynomial of A.

Note that the poles of a matrix of rational functions of s are precisely the roots of the denominator of the matrix.

1.2.3. Exercise. Let V be a subspace of $\mathbb{C}^n$, and let A be such that $A(V) \subset V$. Show that the eigenpolynomial of $A|_V$ divides the eigenpolynomial of A.

For t a real number, consider $f_t(s) = e^{ts}$. $f_t(A)$ is the *matrix exponential* of tA and is denoted by

$$f_t(A) = e^{tA} = \exp(tA).$$

The fundamental property of e^{tA} is the following group property:

$$e^{tA}e^{sA} = e^{(t+s)A}, \qquad t, s \text{ real.}$$

This follows from 1.2.1. Moreover, using either (2.1) or (2.3), one can show that

$$\frac{d}{dt}e^{tA} = Ae^{tA} = e^{tA}A \qquad \text{for all } t.$$

This implies that for any x^0 in $\mathbb{C}^n$, the unique solution $x(t)$ in $\mathbb{C}^n$ of

$$\dot{x} = Ax, \qquad x(0) = x^0$$

is

$$x(t) = e^{tA}x^0.$$

1.2.4. Exercise. Compute e^{tA} for A as in Section 1.1.

One can also use (2.3) to show that $e^A e^B = e^{A+B}$ for any two matrices A, B that commute: $AB = BA$.

1.2.5. Exercise. Give an example of two matrices A, B such that $e^A e^B \neq e^{A+B}$.

1.2.6. Exercise. Show that for any entire function $f(s)$, there are scalars $c_1, \ldots, c_n$ such that

$$f(A) = c_1 I + c_2 A + \cdots + c_n A^{n-1} \qquad (A \text{ is } n \text{ by } n).$$

In particular, for $f_t(s) = e^{ts}$, show that

$$e^{tA} = c_1(t)I + c_2(t)A + \cdots + c_n(t)A^{n-1}$$

for some continuous functions of time, $c_j(t), j = 1, \ldots, n$.

1.2.7. Exercise. Show that the Laplace transform of e^{tA}, $t \geq 0$, is the resolvent $(sI - A)^{-1}$.

Exercise 1.2.7 yields an effective method for computing e^A: simply compute the resolvent, then take the inverse Laplace transform, then set $t = 1$!

We discuss now the concept of stability. An n by n matrix $\bar{A}$ is called *stable* if all the eigenvalues λ of $\bar{A}$ have negative real parts, $\text{Re}(\lambda) < 0$. A polynomial $\bar{q}(s)$ is *stable* if all its roots have negative real parts.

1.2.8. Exercise. Let $\bar{A}$ be stable with all eigenvalues λ satisfying $\text{Re}(\lambda) < -k < 0$. Show that

$$|e^{t\bar{A}}x| \leq |e^{t\bar{A}}||x| \leq \text{constant} \times e^{-kt}, \qquad t \geq 0.$$

(*Hint*: Use (2.4)).

1.2.9. Exercise. Show that $\bar{A}$ is stable if and only if for all x^0 in $\mathbb{C}^n$, the solution $x(t), t \geq 0$, of $\dot{x} = \bar{A}x$, $x(0) = x^0$ in $\mathbb{C}^n$ satisfies $x(t) \to 0$ as $t \uparrow \infty$.

1.2.10. Exercise. Let $\bar{A}$ be in canonical form:

$$\bar{A} = \begin{pmatrix} 0 & 1 & 0 & \cdots & 0 \\ 0 & 0 & 1 & \cdots & 0 \\ 0 & 0 & 0 & \cdots & 0 \\ \cdots\cdots\cdots\cdots\cdots\cdots\cdots \\ 0 & 0 & 0 & \cdots & 1 \\ -q_1 & -q_2 & -q_3 & \cdots & -q_n \end{pmatrix}.$$

Show that $\bar{A}$ is stable iff the polynomial $q(s) = s^n + q_n s^{n-1} + \cdots + q_1$ is stable.

1.2.11. Exercise. Let $\bar{A}$ be a stable matrix. Set

$$L = \int_0^\infty e^{t\bar{A}^*} Q e^{t\bar{A}} \, dt.$$

By Exercise 1.2.8, L is well defined. Show that L is the unique solution of the *Lyapunov equation*

$$\bar{A}^*L + L\bar{A} + Q = 0.$$

An n by n matrix Q is *nonnegative* if Q is self-adjoint, $Q = Q^*$, and for all x, $x^*Qx \geq 0$. Q is *positive* if Q is self-adjoint and for all $x \neq 0$, $x^*Qx > 0$. If Q is self-adjoint, then the eigenvalues of Q are real; if Q is nonnegative, then the eigenvalues are also nonnegative; if Q is positive, then the eigenvalues are also positive.

1.2.12. Exercise. Referring to 1.2.11, show that $Q \to L$ is a bijection between the space of all self-adjoint matrices to itself. Show that $Q \geq 0$ implies $L \geq 0$.

Given two self-adjoint matrices Q_1 and Q_2, we say that $Q_1 \geq Q_2$ if $Q_1 - Q_2$ is nonnegative, $Q_1 - Q_2 \geq 0$. Exercise 1.2.12 then implies that $Q_1 \geq Q_2$ implies that $L_1 \geq L_2$.

1.2.13. Exercise. Let $A(t)$, $0 \leq t \leq T$, be an n by n matrix-valued function of time. Show that

$$\left| \int_0^T A(t)\, dt \right| \leq \int_0^T |A(t)|\, dt.$$

1.2.14. Exercise. Let L_1, L_2 be self-adjoint. Show that $L_1 \geq L_2 > 0$ implies $L_1^{-1} \leq L_2^{-1}$ and $\text{trace}(L_1) \geq \text{trace}(L_2)$.

1.3. Linear Systems

A *linear control system* is a triple (A, B, C), where A is an n by n matrix, B is an n by m matrix, and C is a p by n matrix, all with complex entries (possibly real). m is *the number of inputs*, n is *the dimension*, and p is *the number of outputs*.

A *control* is any function of time $u(t)$, $t \geq 0$, with values in $\mathbb{C}^m$, such that

$$\int_0^T |u(t)|^2\, dt$$

is finite for all T positive.

Given a control $u(t)$, $t \geq 0$, consider the system

$$\dot{x} = Ax + Bu, \qquad x(0) = x^0 \text{ in } \mathbb{C}^n. \qquad (3.1)$$

The *state trajectory* corresponding to the control $u(\cdot)$ is the unique solution $x(t) = x^u(t)$, $t \geq 0$, satisfying (3.1) for all time:

$$x^u(t) = e^{tA}x^0 + \int_0^t e^{(t-s)A}Bu(s)\, ds, \qquad t \geq 0. \qquad (3.2)$$

1.3.1. Exercise. Verify that (3.2) does satisfy (3.1). Also check that (3.1) has a unique solution.

The *output* is the function of time

$$y(t) = y^u(t) = Cx^u(t), \qquad t \geq 0.$$

Thus $y(\cdot)$ takes values in $\mathbb{C}^p$ and

$$y^u(t) = Ce^{tA}x^0 + \int_0^t Ce^{(t-s)A}Bu(s)\,ds, \qquad t \geq 0. \tag{3.3}$$

The p by m matrix-valued function of time

$$Ce^{tA}B, \qquad t \geq 0,$$

is the *weighing pattern* or *impulse response* while (its Laplace transform)

$$G(s) = C(sI - A)^{-1}B \tag{3.4}$$

is the *transfer function*. The map (3.3) $u(\cdot) \to y^u(\cdot)$ is called *the input–output map starting from x^0*. This map is denoted $I/O(x^0)$.

Two systems (A_1, B_1, C_1) and (A_2, B_2, C_2) are said to be *equivalent* if any of the following (equivalent) statements holds:

The transfer functions are the same: $G_1(s) = G_2(s)$.
The weighing patterns are the same: $C_1 e^{tA_1}B_1 = C_2 e^{tA_2}B_2$, $t \geq 0$.
For all $k = 0, 1, 2, \ldots$, $C_1 A_1^k B_1 = C_2 A_2^k B_2$,
The input–output maps *starting at zero* are the same: $I/O_1(0) = I/O_2(0)$.

Note that they then have the same number of inputs and outputs. Their dimensions, however, may be different.

1.3.2. Exercise (*n*th-order integrator). Let $q(s) = s^n + q_n s^{n-1} + \cdots + q_1$ and let $p(s) = p_n s^{n-1} + \cdots + p_1$. Consider the ordinary differential equation

$$x^{(n)} + q_n x^{(n-1)} + \cdots + q_1 x = u,$$

$$y = p_n x^{(n-1)} + \cdots + p_1 x.$$

By setting $x_1 = x$, $x_2 = x'$, $\ldots$, $x_n = x^{(n-1)}$, write this system in the form $\dot{x} = Ax + bu$, $y = cx$, where x is the column vector with entries $x_1, \ldots, x_n$. Compute the eigenpolynomial of A and the transfer function $G(s)$.

Given a p by m matrix of rational functions $G(s)$, one says that a triple (A, B, C) is a *realization* of $G(s)$ if (3.4) holds. For example, if (A, B, C) is a realization of a given $G(s)$, and P is an invertible n by n matrix, then (PAP^{-1}, PB, CP^{-1}) is an equivalent realization.

1.3.3. Exercise. Let T be a positive time and let $u(\cdot)$, $v(\cdot)$ be controls. The *concatenation* of $u(\cdot)$ and $v(\cdot)$ at time T is the control $(u *_T v)(\cdot)$, where

$$(u *_T v)(t) = u(t) \qquad \text{if} \quad t < T$$

$$= v(t - T) \qquad \text{if} \quad t \geq T.$$

Let $x^u(t; x^0)$, $t \geq 0$, be the solution of (3.1), i.e., the state trajectory starting at x^0 at time zero. Show that for $t \geq 0$

$$x^v(t; x^u(T; x^0)) = x^w(t + T; x^0), \qquad w = u *_T v.$$

1.3.4. Exercise. Compute the transfer function of the triple appearing in Section 1.1.

1.3.5. Exercise. Let $\bar{A}$ be stable and suppose

$$\int_0^\infty |v(t)| \, dt < +\infty. \tag{3.5}$$

Let $x^v(\cdot)$ denote the state trajectory corresponding to $(\bar{A}, B)$. Show that $x^v(t) \to 0$ as $t \uparrow \infty$.

1.3.6. Exercise. Referring to 1.3.5, suppose that instead of (3.5) one has

$$\int_0^\infty |v(t)|^2 \, dt < +\infty.$$

Show that there is a positive $L > 0$ and real numbers $\bar{a} < 0$, $b > 0$, such that $x^v(t)*Lx^v(t) \leq s(t)$, $t \geq 0$, where $s(\cdot)$ is the solution of the scalar equation

$$\dot{s} = \bar{a}s + b|u|^2, \qquad s(0) = x^{0*}Lx^0.$$

Conclude that $x^v(t) \to 0$ as $t \uparrow \infty$. (*Hint*: Use the Lyapunov equation.)

1.3.7. Exercise. Extending 1.3.5 and 1.3.6, suppose that instead of (3.5) one has

$$\int_0^\infty |v(t)|^r \, dt < +\infty, \tag{3.6}$$

for some $r \geq 1$. Show that $x^v(t) \to 0$ as $t \uparrow \infty$. (*Hint*: Look up the Holder inequality, and use the fact that $e^{\bar{A}t}$, $t \geq 0$, is in $L^{r'}$ where $r^{-1} + r'^{-1} = 1$.)

1.3.8. Exercise. Extending 1.3.7, suppose that $v(\cdot)$ satisfies (3.6). Show that

$$\int_0^\infty |x^v(t)|^r \, dt \leq \text{constant} \times \left(\int_0^\infty |v(t)|^r \, dt + |x^0|^r \right).$$

(*Hint*: Look up Young's inequality, i.e., the fact that the convolution of L^r with L^1 is back in L^r.)

1.4. Controllability and Observability

Let

$$\dot{x} = Ax + Bu, \qquad x(0) = x^0 \text{ in } \mathbb{C}^n, \tag{4.1}$$

$$y = Cx \tag{4.2}$$

be a linear system. A state x is said to be *controllable from zero* in time $T > 0$ if there is a control $u(\cdot)$ such that the state trajectory $x^u(\cdot)$ corresponding to $u(\cdot)$ starting at the origin $x^u(0) = 0$ passes through x at time T: $x^u(T) = x$. One says that $u(\cdot)$ "steers" the system from the origin to the state x in time T. Let V^T be the set of all such x. The system is said to be *controllable* if $V^T = \mathbb{C}^n$. A pair of matrices (A, B) is *controllable* if the system (4.1) is so.

1.4.1. Proposition. *V^T is equal to the range space of the matrix $(B, AB, \ldots, A^{n-1}B)$. Thus V^T is independent of T. Also (4.1) is controllable iff the rank of the matrix $(B, AB, \ldots, A^{n-1}B)$ is n.*

PROOF. According to (3.2) x is in V^T iff

$$x = \int_0^T e^{(T-t)A} Bu(t)\, dt \tag{4.3}$$

for some $u(\cdot)$. Thus V^T is a linear subspace of $\mathbb{C}^n$. By Exercise 1.2.6, x is in V^T iff for some continuous functions of time $c_j, j = 1, \ldots, n,$

$$x = \sum_{j=1}^{n} A^{j-1} Bu_j, \tag{4.4}$$

where

$$u_j = \int_0^T c_j(T - t)u(t)\, dt, \qquad j = 1, \ldots, n.$$

Thus (4.4) shows that if x is in V^T, then x is in the range space of the stated matrix. Now we show that if c is perpendicular to V^T, then c is perpendicular to the range space of the stated matrix. This will imply the result. So let c be a row vector perpendicular to V^T. By (4.3) one has

$$\int_0^T ce^{(T-t)A} Bu(t)\, dt = 0$$

for all $u(\cdot)$. Thus the weighing pattern $ce^{tA}B$ vanishes identically, $0 \le t \le T$. Differentiating $j - 1$ times and setting $t = 0$, one arrives at $cA^{j-1}B = 0$ for $j = 1, \ldots, n$; thus c is perpendicular to the range space of $(B, AB, \ldots, A^{n-1}B)$.   □

1.4.2. Exercise. Let A, $B = (b_1, b_2)$ be as in Section 1.1. Which of the pairs (A, B), (A, b_1), and (A, b_2) are controllable?

1.4.3. Exercise. Let $m = 1, p = 1$, and suppose (A_1, b_1) is controllable. Exhibit an n by n invertible matrix P satisfying

$$P^{-1} A_1 P = A, \qquad Pb = b_1, \qquad c = c_1 P,$$

where A, b, and c are as in Exercise 1.3.2, for some choice of polynomials $p(s)$ and $q(s)$. What is the relation of these polynomials to the transfer function of (A_1, b_1, c_1)?

Given a pair (A, B), let V_T be the set of all x^0 in $\mathbb{C}^n$ that are *controllable to zero in time T*. This means there is a control $u(\cdot)$ such that $x^u(0) = x^0$ and $x^u(T) = 0$.

1.4.4. Exercise. Show that $e^{TA}(V_T) = V^T$ and so (A, B) is controllable iff $V_T = V^T = \mathbb{C}^n$.

A state x^0 is said to be *observable from zero in time* $T > 0$ if there is a control $u(\cdot)$ that *distinguishes* between x^0 and the origin: The output $y^u(t), 0 \le t \le T$, of (4.1), (4.2) starting at x^0, is *not* equal to the output $y_0^u(t), 0 \le t \le T$, of (4.1), (4.2), starting at the origin. Appealing to (3.3), we see that this happens iff $Ce^{tA}x^0$ is *not* identically zero for $0 \le t \le T$. Let W^T be the set of all x^0 that are *not* observable from zero in time T. Thus x^0 is in W^T iff

$$Ce^{tA}x^0 = 0, \qquad 0 \le t \le T. \tag{4.5}$$

The system (4.1), (4.2) is *observable* if $W^T = \{0\}$, i.e., all nonzero states are observable. A pair (A, C) is *observable* if the system (4.1), (4.2) is so. Note that this is a property of the matrices A and C only.

1.4.5. Proposition. W^T *is equal to the null-space of the matrix*

$$\begin{pmatrix} C \\ CA \\ \vdots \\ CA^{n-1} \end{pmatrix}. \tag{4.6}$$

Thus W^T is independent of T. Also the system (4.1), (4.2) is observable iff the rank of (4.6) is n.

PROOF. If x^0 is in W^T, then $Ce^{tA}x^0 = 0, 0 \le t \le T$. By differentiating $j - 1$ times, and setting $t = 0$, we have $CA^{j-1}x^0 = 0, j = 1, \ldots, n$. Thus x^0 is in the null-space of (4.6). Conversely, using Exercise 1.2.6, any x^0 in the null-space of (4.6) is in W^T. □

1.4.6. Exercise. (A, C) is observable iff $\text{I/O}(x_1^0) = \text{I/O}(x_2^0)$ implies $x_1^0 = x_2^0$.

1.4.7. Exercise. (A, C) is observable iff $Ce^{tA}x = 0, t \ge 0$, implies $x = 0$.

1.4.8. Exercise. Let $\bar{A}$ be a stable matrix and suppose that $(\bar{A}, C)$ is an observable pair. Let $Q = C^*C$. Show that the solution L of the Lyapunov equation (Exercise 1.2.11) is positive, $L > 0$ (Exercise 1.2.12).

1.4.9. Exercise. Let F be any matrix. Show that (A, B) is controllable if and only if $(A - BF, B)$ is controllable. Do this problem in two different ways.

1.4.10. Exercise. Show that (A, B) is controllable iff (A^*, B^*) is observable, and that (A, C) is observable iff (A^*, C^*) is controllable.

1.4.11. Exercise. Show that (A_1, B_1, C_1) and (A_2, B_2, C_2) are equivalent iff (A_1^*, C_1^*, B_1^*) and (A_2^*, C_2^*, B_2^*) are equivalent.

1.4.12. Exercise. If (A, B, C) is equivalent to (A^*, C^*, B^*) then what can you say about $G(s)$? What can you say about (A, B, C)?

1.4.13. Exercise. Let $m = 1$ and suppose that (A, b) is controllable. Show that the minimal polynomial of A and the eigenpolynomial of A agree.

1.5. Minimality

A triple (or a system) (A, B, C) is *minimal* if (A, B) is controllable and (A, C) is observable.

Consider two linear systems (A, B, C) and $(\tilde{A}, \tilde{B}, \tilde{C})$ of dimension n and $\tilde{n}$, respectively, both having the same number of inputs and outputs. Suppose that they are equivalent. Recall that this means that their corresponding I/O maps agree, when both systems are started at the origin (Section 1.3). Let V^T and $I/O(x)$ denote the controllability subspace and the I/O map corresponding to (A, B, C), and let the corresponding objects for $(\tilde{A}, \tilde{B}, \tilde{C})$ be $\tilde{V}^T$ and $\widetilde{I/O}(x)$.

1.5.1. Theorem. *Let (A, C) be observable. Then there is a linear map P of $\tilde{V}^T$ onto V^T such that for all $\tilde{x}$ in $\tilde{V}^T$,*

$$\widetilde{I/O}(\tilde{x}) = I/O(x), \qquad P\tilde{x} = x. \tag{5.1}$$

The map P is uniquely determined by (5.1). Moreover, if in addition (A, B) is controllable, then $\tilde{n} \geq n$. Thus among all systems equivalent to a given system, the minimal ones have the least dimension.

PROOF. Let $x^u(\cdot; x)$, $y^u(\cdot; x)$ denote the state trajectory and the output of the system (A, B, C) started at x, and let $\tilde{x}^u(\cdot; x)$, $\tilde{y}^u(\cdot; x)$ denote the corresponding objects for $(\tilde{A}, \tilde{B}, \tilde{C})$. Fix $T > 0$.

Let $\tilde{x}$ be in $\tilde{V}^T$. Choose any $u(\cdot)$ such that $\tilde{x} = \tilde{x}^u(T; 0)$. Define P by setting $x = P\tilde{x} = x^u(T; 0)$. We have to show that P is well defined. Assume that $u(t) = 0$ for $t \geq T$. (This does not change x and $\tilde{x}$.)

Suppose now that $v(\cdot)$ is another control such that $\tilde{x}^u(T; 0) = \tilde{x}^v(T; 0) = \tilde{x}$. We have to show that $x_1 = x^u(T; 0)$ and $x_2 = x^v(T; 0)$ are equal. Assume that $v(\cdot)$ also vanishes after time T. Then (Exercise 1.3.3) one has $\tilde{x}^u(t; 0) = \tilde{x}^v(t; 0)$ for $t \geq T$. Thus $\tilde{y}^u(t; 0) = \tilde{y}^v(t; 0)$ for $t \geq T$. Since the systems are equivalent, this implies $y^u(t; 0) = y^v(t; 0)$ for $t \geq T$. By Exercise 1.3.3 again

$$Ce^{tA}x_1 = Cx^0(t; x_1) = Cx^0(t; x^u(T; 0)) = Cx^u(t + T; 0) = y^u(t + T; 0)$$

and similarly

$$Ce^{tA}x_2 = y^v(t + T; 0)$$

for all $t \geq 0$. Thus $Ce^{tA}(x_1 - x_2) = 0$ for $t \geq 0$. By Exercise 1.4.7 this implies $x_1 = x_2$. Thus P is well defined.

Now if $\tilde{x}_1$ and $\tilde{x}_2$ are in $\tilde{V}^T$ and $\tilde{x}_j = \tilde{x}^{u_j}(T; 0)$, $j = 1, 2$, then $a_1\tilde{x}_1 + a_2\tilde{x}_2$ is also in $\tilde{V}^T$ and is equal to $\tilde{x}^v(T; 0)$, where $v = a_1u_1 + a_2u_2$. Thus

$$
\begin{aligned}
P(a_1\tilde{x}_1 + a_2\tilde{x}_2) &= P(\tilde{x}^v(T; 0)) \\
&= x^v(T; 0) \\
&= a_1x^{u_1}(T; 0) + a_2x^{u_2}(T; 0) \\
&= a_1x_1 + a_2x_2 \\
&= a_1P(\tilde{x}_1) + a_2P(\tilde{x}_2).
\end{aligned}
$$

This shows that P is linear. If x is in V^T, then $x = x^u(T; 0)$ for some control $u(\cdot)$. Set $\tilde{x} = \tilde{x}^u(T; 0)$. Then $P\tilde{x} = x$. Thus P is onto V^T. Now let $\tilde{x}$ be in $\tilde{V}^T$ and let $u(\cdot)$ be any control. Let $v(\cdot)$ be a control such that $\tilde{x} = \tilde{x}^v(T; 0)$. Then let $x = P\tilde{x} = x^v(T; 0)$. Then for $t \geq 0$

$$
\begin{aligned}
\tilde{y}^u(t; \tilde{x}) &= C\tilde{x}^u(t; \tilde{x}) \\
&= C\tilde{x}^u(t; \tilde{x}^v(T; 0)) \\
&= C\tilde{x}^w(t + T; 0) \qquad (w = u *_T v, \text{ Exercise 1.3.3}) \\
&= \tilde{y}^w(t + T; 0) \\
&= y^w(t + T; 0) \qquad \text{(equivalent systems)} \\
&= Cx^w(t + T; 0) \\
&= Cx^u(t; x^v(T; 0)) \qquad \text{(Exercise 1.3.3)} \\
&= Cx^u(t; x) \\
&= y^u(t; x).
\end{aligned}
$$

This proves (5.1). Now suppose that in addition (A, B) is controllable. Then $V^T = \mathbb{C}^n$. Thus $\tilde{n} = \dim(\mathbb{C}^{\tilde{n}}) \geq \dim(\tilde{V}^T) \geq \dim(V^T) = \dim(\mathbb{C}^n) = n$. This shows that $\tilde{n} \geq n$. Now suppose that there are two linear maps P_1 and P_2 satisfying (5.1). Then for all $\tilde{x}$ in $\tilde{V}^T$,

$$I/O(P_1\tilde{x}) = \widetilde{I/O}(\tilde{x}) = I/O(P_2\tilde{x}).$$

Then observability of (A, C) implies (Exercise 1.4.6) that $P_1\tilde{x} = P_2\tilde{x}$. This shows that P is unique, and completes the proof. $\qquad \qquad \square$

1.5.2. Corollary. *Let $(\tilde{A}, \tilde{B}, \tilde{C})$ be controllable on $\mathbb{C}^{\tilde{n}}$ and let (A, B, C) be observable on $\mathbb{C}^n$. Assume that $(\tilde{A}, \tilde{B}, \tilde{C})$ and (A, B, C) are equivalent. Then there is a unique n by $\tilde{n}$ matrix P such that*

$$AP = P\tilde{A}, \qquad P\tilde{B} = B, \qquad \tilde{C} = CP. \tag{5.2}$$

PROOF. Clearly, any P that satisfies (5.2) satisfies (5.1). Thus such a P is uniquely determined. Now, by 1.5.1, there is a linear map $P: \tilde{V}^T \to V^T \subset \mathbb{C}^n$ such that (5.1) holds. Since $(\tilde{A}, \tilde{B}, \tilde{C})$ is controllable, $\mathbb{C}^{\tilde{n}} = \tilde{V}^T$ and so $P: \mathbb{C}^{\tilde{n}} \to \mathbb{C}^n$ is (multiplication by) an n by $\tilde{n}$ matrix. Appealing to (5.1) with $u(\cdot) = 0$, one has $\tilde{C}e^{t\tilde{A}}\tilde{x} = Ce^{tA}P\tilde{x}$, $t \geq 0$. Setting $t = 0$ yields $\tilde{C} = CP$. Now

$$Ce^{tA}AP = \frac{d}{dt}Ce^{tA}P = \frac{d}{dt}\tilde{C}e^{t\tilde{A}} = \tilde{C}e^{t\tilde{A}}\tilde{A} = Ce^{tA}P\tilde{A}, \qquad t \geq 0.$$

Now Exercise 1.4.7 implies that $AP = P\tilde{A}$. Similarly,

$$Ce^{tA}B = \tilde{C}e^{t\tilde{A}}\tilde{B} = Ce^{tA}P\tilde{B}, \qquad t \geq 0.$$

Using Exercise 1.4.7 again yields $P\tilde{B} = B$. $\qquad\qquad\qquad\qquad\qquad\qquad\square$

1.5.3. Corollary. *Let (A_j, B_j, C_j), $j = 1, 2$, be two equivalent minimal systems. Then there is a unique invertible matrix P such that*

$$PA_1P^{-1} = A_2, \qquad PB_1 = B_2, \qquad C_1 = C_2P.$$

PROOF. By 1.5.1, the dimensions of the two systems must be the same, and hence P is an invertible matrix, since P is onto $\mathbb{C}^n$. The rest follows from 1.5.2. $\qquad\square$

1.5.4. Exercise. Let J be the $2n$ by $2n$ matrix $\begin{pmatrix} 0 & I \\ -I & 0 \end{pmatrix}$. A system (A, B, C) of dimension $2n$ is called *Hamiltonian* if

$$JA = -A^*J \quad \text{and} \quad JB = -C^*.$$

A transfer function $G(s)$ is *Hamiltonian* if $G(-s^*)^* = G(s)$. Show that (A, B, C) Hamiltonian implies $G(s)$ Hamiltonian.

Note that the poles of $G(s) = C(sI - A)^{-1}B$ are also the poles of the resolvent $(sI - A)^{-1}$. Thus the poles of $G(s)$ are eigenvalues of A, though not necessarily all of them.

1.5.5. Exercise. If (A, B, C) is minimal, then the denominator of $G(s)$ (see 1.2.2) is the minimal polynomial of A.

A p by m matrix $G(s)$ of rational functions of s is *stable* if all its poles lie in the left-half plane. Conclude that if (A, B, C) is a minimal realization of $G(s)$, then A is stable if and only if $G(s)$ is stable.

Let λ be an eigenvalue of A, $Ax = \lambda x$ with $x \neq 0$. We say that λ is an *observable* eigenvalue if $Cx \neq 0$, which happens iff x is not in W^T.

1.5.6. Exercise. Let (A, B) be controllable. If λ is an observable eigenvalue of A, then λ is a pole of $G(s) = C(sI - A)^{-1}B$.

1.5.7. Exercise. Show that (A, C) is observable if and only if all eigenvalues of A are observable.

1.5.8. Exercise. Referring to 1.5.4, show that if λ is an eigenvalue of A, then so is $-\lambda^*$; also show that (A, B) is controllable iff (A, C) is observable. Show that $q(s) = \det(sI - A)$ satisfies $q(-s^*)^* = q(s)$.

1.6. Realizability

Given a triple (A, B, C) the transfer function

$$G(s) = C(sI - A)^{-1}B \tag{6.1}$$

is a p by m matrix with entries that are rational functions of s. As $s \to \infty$, $G(s) \to 0$. Thus the entries are *proper* rational functions. We consider the following inverse problem: when does a p by m matrix of proper rational functions arise as a transfer function? We shall see below that this is always the case. It is then reasonable to ask for an algorithm that produces a realization (A, B, C) of the given matrix $G(s)$. This is done in the following proposition. The realization (A, B, C) constructed here is called the *standard controllable realization*.

1.6.1. Proposition. *Let $G(s)$ be a p by m matrix of proper rational functions of s. Then there is a triple (A, B, C) satisfying (6.1).*

PROOF. Write $G(s)$ as $P(s)/q(s)$, where $P(s)$ is a matrix with polynomial entries and

$$q(s) = s^r + q_r s^{r-1} + \cdots + q_1.$$

We construct a realization of dimension $n = mr$. Let

$$P(s) = P_r s^{r-1} + \cdots + P_1.$$

Letting I_m and 0_m denote the identity matrix and the zero matrix of size m by m, we set

$$
A = \begin{pmatrix}
0_m & I_m & \cdots & 0_m \\
0_m & 0_m & \cdots & 0_m \\
 & & \cdots & \\
0_m & 0_m & \cdots & I_m \\
-q_1 I_m & -q_2 I_m & \cdots & -q_r I_m
\end{pmatrix}, \quad
B = \begin{pmatrix}
0_m \\
0_m \\
\vdots \\
0_m \\
I_m
\end{pmatrix}, \quad
C = (P_1, P_2, \ldots, P_r).
$$

Now direct computation yields

$$(sI - A)^{-1}B = \frac{1}{q(s)}\begin{pmatrix} I_m \\ sI_m \\ \vdots \\ s^{r-1}I_m \end{pmatrix}; \tag{6.2}$$

indeed, multiplying the right-hand side of (6.2) by $(sI - A)$ yields the matrix B. Hence

$$C(sI - A)^{-1}B = \frac{1}{q(s)}(P_1, \ldots, P_r)\begin{pmatrix} I_m \\ sI_m \\ \vdots \\ s^{r-1}I_m \end{pmatrix}$$

$$= \frac{P_r s^{r-1} + \cdots + P_1}{s^r + q_r s^{r-1} + \cdots + q_1} = \frac{P(s)}{q(s)} = G(s).$$

This completes the proof. (Compare with Exercises 1.3.2 and 1.4.3.) $\square$

1.6.2. Corollary. *If $G(s)$ has coefficients in a field $k \subset \mathbb{C}$, then $G(s)$ has a realization (A, B, C) with entries in k.*

PROOF. This follows immediately from the construction in 1.6.1. $\square$

In particular, real transfer functions are always realizable by real matrices (A, B, C).

The realization constructed in 1.6.1 need not be minimal. Indeed, by multiplying the numerator and denominator of $G(s)$ by a power of s, one can increase the degree of $q(s)$ without changing $G(s)$, and thus increase the state dimension of the exhibited realization. Even if $q(s)$ were chosen to be the least common multiple of the denominators of the entries of $G(s)$, the realization appearing in 1.6.1 need not be minimal. It turns out however that it is always controllable. Indeed, with (A, B) as in 1.6.1, one can verify that $\text{rank}(B, AB, \ldots, A^{r-1}B) = mr = n$.

1.6.3. Exercise. Do it.

1.6.4. Exercise. (See 1.2.2 and 1.5.5.) Let $m = 1$. Write $G(s) = P(s)/q(s)$ with $q(s)$ the denominator of $G(s)$. Show that the standard controllable realization (A, B, C) exhibited in 1.6.1 is minimal.

Thus if $m = 1$, one can construct a canonical minimal realization. If $p = 1$, then, by using 1.4.10, one can also construct a canonical minimal realization. However, if $\min(m, p) > 1$, then no such canonical minimal realization can be constructed.

We now turn to the problem of constructing an observable realization equivalent to a given one.

1.6.5. Proposition. *Let* $(\tilde{A}, \tilde{B}, \tilde{C})$ *have dimension* $\tilde{n}$. *Then there is an equivalent observable realization* (A, B, C) *on* $\mathbb{C}^n$ *for some* $n \leq \tilde{n}$. *If* $(\tilde{A}, \tilde{B}, \tilde{C})$ *is controllable, then* (A, B, C) *can also be chosen to be controllable and hence minimal.*

PROOF. According to 1.5.2, *if* there were such an (A, B, C) and *if* $(\tilde{A}, \tilde{B}, \tilde{C})$ were controllable, then there would exist an n by $\tilde{n}$ matrix P such that (5.2) held. Without assuming that $(\tilde{A}, \tilde{B}, \tilde{C})$ is controllable, let us now construct such a P. Consider the $p\tilde{n}$ by $\tilde{n}$ matrix

$$\begin{pmatrix} \tilde{C} \\ \tilde{C}\tilde{A} \\ \vdots \\ \tilde{C}\tilde{A}^{\tilde{n}-1} \end{pmatrix}. \tag{6.3}$$

Let the rank of this matrix be $n \leq \tilde{n}$. Choose an n by $\tilde{n}$ submatrix P of (6.3) having rank n. Choose an $\tilde{n}$ by n matrix Q such that $PQ = I_n$. Set

$$A = P\tilde{A}Q, \qquad B = P\tilde{B}, \qquad C = \tilde{C}Q.$$

We claim that (A, B, C) is the required realization.

Since P has rank n, the same as the rank of (6.3), all rows of (6.3) can be expressed as a linear combination of P:

$$\tilde{C}\tilde{A}^{j-1} = Z_j P \tag{6.4}$$

for some p by n matrix Z_j, for all $j = 1, \ldots, n$. By Cayley–Hamilton, (6.4) will hold for some Z_j, for *all* $j = 1, 2, \ldots$. We use this to show that if a vector v in $\mathbb{C}^{\tilde{n}}$ satisfies $Pv = 0$, then we also have $P\tilde{A}v = 0$. Indeed, $Pv = 0$ implies, by (6.4), that $\tilde{C}\tilde{A}^{j-1}v = 0$ for all j. Hence $\tilde{C}\tilde{A}^{j-1}(\tilde{A}v) = 0$ for all j. Hence $P\tilde{A}v = 0$. Now for any $\tilde{x}$ in $\mathbb{C}^{\tilde{n}}$, set $v = QP\tilde{x} - \tilde{x}$. Then $Pv = P(QP\tilde{x} - \tilde{x}) = PQP\tilde{x} - P\tilde{x} = P\tilde{x} - P\tilde{x} = 0$. Thus $P\tilde{A}v = 0$. Hence

$$(AP - P\tilde{A})\tilde{x} = P\tilde{A}QP\tilde{x} - P\tilde{A}\tilde{x} = P\tilde{A}(QP\tilde{x} - \tilde{x}) = P\tilde{A}v = 0.$$

Similarly,

$$(CP - \tilde{C})\tilde{x} = \tilde{C}QP\tilde{x} - \tilde{C}\tilde{x} = \tilde{C}v = Z_1 Pv = 0.$$

Thus (5.2) is verified. It is easy to check that (5.2) implies that $\tilde{C}\tilde{A}^k\tilde{B} = CA^kB$ for $k = 0, 1, \ldots$ and so (A, B, C) is equivalent to $(\tilde{A}, \tilde{B}, \tilde{C})$. Now

$$n \geq \mathrm{rank}\begin{pmatrix} C \\ CA \\ \vdots \\ CA^{n-1} \end{pmatrix} = \mathrm{rank}\begin{pmatrix} C \\ CA \\ \vdots \\ CA^{\tilde{n}-1} \end{pmatrix} \qquad \text{(by Cayley–Hamilton)}$$

$$\geq \mathrm{rank}\begin{pmatrix} C \\ CA \\ \vdots \\ CA^{\tilde{n}-1} \end{pmatrix} P$$

$$= \operatorname{rank} \begin{pmatrix} \tilde{C} \\ \tilde{C}\tilde{A} \\ \vdots \\ \tilde{C}\tilde{A}^{\tilde{n}-1} \end{pmatrix} = n.$$

Thus (A, C) is observable. Now assume that $(\tilde{A}, \tilde{B}, \tilde{C})$ is controllable.

To check that (A, B) is controllable, note that since $(\tilde{A}, \tilde{B})$ is so, $(\tilde{B}, \tilde{A}\tilde{B}, \ldots, \tilde{A}^{\tilde{n}-1}\tilde{B})$: $\mathbb{C}^{\tilde{n}m} \to \mathbb{C}^{\tilde{n}}$ is onto. Since $P: \mathbb{C}^{\tilde{n}} \to \mathbb{C}^n$ is also onto, we have that $(B, AB, \ldots, A^{\tilde{n}-1}B) = P(\tilde{B}, \tilde{A}\tilde{B}, \ldots, \tilde{A}^{\tilde{n}-1}\tilde{B})$ is also onto. Thus rank $(B, AB, \ldots, A^{n-1}B) = \operatorname{rank}(B, AB, \ldots, A^{\tilde{n}-1}B) = n$. This completes the proof. $\qquad\square$

Proposition 1.6.1 and 1.6.5 taken together yield an algorithm that produces a minimal realization of a given transfer function. There are other algorithms.

1.6.6. Exercise. Let A, b_1, and C be as in Section 1.1. Compute a minimal realization of $G(s) = C(sI - A)^{-1}b_1$.

The following exercise provides us with the "dual" of 1.6.1.

1.6.7. Exercise. Let $G(s)$ be a p by m matrix of proper rational functions. Then there is a realization (A, B, C) of $G(s)$ that is observable. Do this without appealing to 1.6.5. (*Hint*: Look at 1.4.10.)

Now for the dual version of 1.6.5.

1.6.8. Proposition. *Let (A, B, C) have dimension n. Then there is an equivalent controllable realization $(\tilde{A}, \tilde{B}, \tilde{C})$ on $\mathbb{C}^{\tilde{n}}$ for some $\tilde{n} \leq n$. If (A, B, C) is observable, then $(\tilde{A}, \tilde{B}, \tilde{C})$ can also be chosen to be observable and hence minimal. WARNING: This time we have $\tilde{n} \leq n$.*

PROOF. Use Exercises 1.4.10 and 1.4.11 together with 1.6.5. $\qquad\square$

We conclude with a "partial-fractions" algorithm.

1.6.9. Exercise. Let $G(s)$ be a sum of terms $G_k(s)$, $1 \leq k \leq N$, where each $G_k(s) = G_k/(s - \lambda_k)$. Here $\lambda_1, \ldots, \lambda_N$ are *distinct* complex numbers. Show that

$$\dot{x}_k = \lambda_k x_k + B_k u, \qquad x_k(0) = x_k^0 \text{ in } \mathbb{C}^{n_k}, \qquad y_k = C_k x_k,$$

$$y = y_1 + \cdots + y_N, \qquad G_k = C_k B_k, \qquad \operatorname{rank}(G_k) = n_k = \operatorname{rank}(B_k) = \operatorname{rank}(C_k)$$

is a minimal realization of $G(s)$. Thus the dimension of $G(s)$ is $n_1 + \cdots + n_N$.

1.7. Notes and References

The material in this chapter is classical in the sense that the results appearing here were all well known by the late 1960s. The point of view taken here is based largely on Brockett's book [1.1]. The example of Section 1.1 is taken from there. In addition the problems appearing there are an excellent source of motivating and illustrative material. Section 1.2 is meant to be a review of linear algebra; the problems developed there are used throughout Chapters 1 and 2. Except for "eigenpolynomial" the terminology of Section 1.2 is standard. Background material may be found in [1.4]. Holder's inequality may be found in [1.10] and *Young's inequality* in [1.11].

An early paper discussing controllability is Pontrjagin [1.9]. Observability was introduced by Kalman [1.5], [1.6] who used it in the solution of the LQ regulator. Minimality is developed in [1.6] and the paper of Gilbert [1.2]. There is a construction dual to that of 1.6.1, the "standard observable realization." This appears on p. 107 of [1.1]. The nonexistence of canonical forms when $\min(m, p) > 1$ can be found in Hazewinkel [1.3]. An introduction to nonlinear control is the NASA Ames Conference Proceedings [1.8].

[1.1]   R. W. Brockett, *Finite Dimensional Linear Systems*, Wiley, New York, 1970.
[1.2]   E. G. Gilbert, "Controllability and Observability in Multivariable Control Systems," *SIAM J. Control Optim.*, **1** (1963), 128–151.
[1.3]   M. Hazewinkel, "Moduli and Canonical Forms for Linear Dynamical Systems," appears in [1.8].
[1.4]   M. W. Hirsch and S. Smale, *Differential Equations, Dynamical Systems, and Linear Algebra*, Academic Press, New York, 1974.
[1.5]   R. E. Kalman, "Canonical Structure of Linear Dynamical Systems," *Proc. Nat. Acad. Sci. U.S.A*, **48** (1962), 596–600.
[1.6]   ——, "Mathematical Description of Linear Dynamical Systems," *SIAM J. Control Optim.*, **1** (1963), 152–192.
[1.7]   ——, Y. C. Ho, and K. S. Narendra, "Controllability of Linear Dynamical Systems," in *Contributions to the Theory of Differential Equations*, Vol. I. Interscience, New York, 1963.
[1.8]   C. Martin and R. Hermann (eds.), *Proceedings of the* 1976 *Ames Research Center* (*NASA*) *Conference on Geometric Control Theory*, Mathematical Science Press, Brookline, MA, 1977.
[1.9]   L. S. Pontrjagin, "Optimal Regulation Processes," *Uspekhi Mat. Nauk.* (*N.S.*), **14** (1959), 3–20. (*AMS Translations* (*Series* 2), **18** (1961), 321–340.)
[1.10] W. Rudin, *Real and Complex Analysis*, McGraw-Hill, New York, 1970.
[1.11] E. M. Stein and G. Weiss, *Fourier Analysis in Euclidean Space*, Princeton University Press, Princeton, NJ, 1971.

CHAPTER 2
The *LQ* Regulator

2.1. Stabilization

Exercise 1.2.8, 1.2.9, 1.2.10, and 1.2.11 should be worked out before starting this section.

Consider the system

$$\dot{x} = Ax + Bu, \qquad x(0) = x^0 \text{ in } \mathbb{C}^n. \tag{1.1}$$

We desire to stabilize (1.1), i.e., to find a control $u(\cdot)$ such that the solution $x^u(\cdot)$ of (1.1) converges to zero as $t \uparrow \infty$, $x^u(t) \to 0$ as $t \uparrow \infty$.

One approach to this problem is to seek an m by n matrix F such that the *feedback law* $u = -Fx$ yields the desired control $u(\cdot)$. This means that $u(\cdot)$ is the control satisfying

$$u(t) = -Fx^u(t), \qquad t \geq 0. \tag{1.2}$$

2.1.1. Proposition. *Let F be an m by n matrix. Then given an initial state x^0, there is a unique control $u^{\#}(\cdot)$ satisfying* (1.2).

PROOF. Let $x^{\#}(t)$, $t \geq 0$, be the solution of

$$\dot{x} = (A - BF)x, \qquad x(0) = x^0 \text{ in } \mathbb{C}^n. \tag{1.3}$$

Set $u^{\#}(t) = -Fx^{\#}(t)$, $t \geq 0$. Then $x^{\#}(\cdot)$ satisfies (1.1) with $u(\cdot) = u^{\#}(\cdot)$. Thus $x^{\#}(\cdot) = x^{u^{\#}}(\cdot)$, and so $u^{\#}(\cdot)$ satisfies (1.2). Conversely, if $u(\cdot)$ is any control satisfying (1.2), then $x^u(\cdot)$ satisfies (1.3) and hence $x^u(\cdot) = x^{\#}(\cdot)$. By (1.2), again, $u = -Fx^u = -Fx^{\#} = u^{\#}$. This shows that $u^{\#}(\cdot)$ is unique, and completes the proof. $\qquad\square$

Let us make things precise. We say that *a control $u(\cdot)$ is stabilizing at x^0* if the solution of (1.1) satisfies $x^u(t) \to 0$ as $t \uparrow \infty$. We say that a feedback law

$u = -Fx$ is *stabilizing* if for all initial states x^0, the control $u(\cdot)$ satisfying (1.1) and (1.2) is stabilizing at x^0. Thus possession of a stabilizing feedback law enables one to produce, for each x^0, a control that is stabilizing at x^0.

We now pose the question: given A and B how do we find an F such that the feedback law (1.2) is stabilizing? Equation (1.3), together with Exercise 1.2.9, gives the answer: choose F such that the matrix $\bar{A} = A - BF$ is stable. This however does not really solve the problem but merely isolates it:

Given a pair (A, B) find a matrix F such that $\bar{A} = A - BF$ is stable.

We hasten to point out that this may not be possible. Indeed, take $B = 0$. Then no choice of F can affect $\bar{A} = A$. On the other hand, if $m = n$, and B is invertible, then choosing $F = B^{-1}(\bar{A} - A)$ yields the fact that $\bar{A}$ can be prescribed arbitrarily. Conversely, if $\bar{A}$ can be prescribed arbitrarily, then B has to be invertible, or at least rank n, if $m \geq n$.

A pair (A, B) for which a stabilizing feedback F can be found is called *stabilizable*. One of the results of this chapter is that controllability implies stabilizability.

2.1.2. Exercise. Does stabilizability imply controllability?

Let us see what stabilizability entails in the case of a single input, $m = 1$, and (A, b) is in standard controllable form

$$
A = \begin{pmatrix} 0 & 1 & \cdots & 0 \\ 0 & 0 & \cdots & 0 \\ & & \cdots & \\ 0 & 0 & \cdots & 1 \\ -q_1 & -q_2 & \cdots & -q_n \end{pmatrix}, \qquad b = \begin{pmatrix} 0 \\ 0 \\ \vdots \\ 0 \\ 1 \end{pmatrix}.
$$

Then F is a 1 by n row vector $f = (f_1, \ldots, f_n)$ and

$$
\bar{A} = A - bf = \begin{pmatrix} 0 & 1 & \cdots & 0 \\ 0 & 0 & \cdots & 0 \\ & & \cdots & \\ 0 & 0 & \cdots & 1 \\ -\bar{q}_1 & -\bar{q}_2 & \cdots & -\bar{q}_n \end{pmatrix} \qquad \text{with} \quad \bar{q}_j = q_j + f_j, \quad j = 1, \ldots, n.
$$

Thus $\bar{A}$ is also in standard form and hence (Exercise 1.2.10) $\bar{A}$ is stable iff $\bar{q}(s) = s^n + \bar{q}_n s^{n-1} + \cdots + \bar{q}_1$ is a stable polynomial. The problem thus reduces to the following: given a polynomial $q(s) = s^n + q_n s^{n-1} + \cdots + q_1$ find a polynomial $f(s) = f_n s^{n-1} + \cdots + f_1$ such that $\bar{q}(s) = q(s) + f(s)$ is stable. But this can always be done!

2.1.3. Proposition. *Given a polynomial $q(s)$ as above and any complex numbers $\lambda_1, \ldots, \lambda_n$ (not necessarily distinct) there is a polynomial $f(s)$ as above such that*

$$
\bar{q}(s) = q(s) + f(s) = (s - \lambda_1)(s - \lambda_2) \ldots (s - \lambda_n),
$$

i.e., $\bar{q}(s)$ has roots at arbitrarily prescribed locations $\lambda_1, \ldots, \lambda_n$.

PROOF. This is clear. Simply choose $f_j = -q_j + \bar{q}_j$ and choose $\bar{q}_j$ such that the roots are where you want them. □

Thus for (A, b) in standard controllable form, not only does a stabilizing feedback exist but one can also, by an appropriate choice of f, arbitrarily arrange the eigenvalues of $\bar{A}$.

2.1.4. Exercise. Let $m = 1$, and let (A, b) be a controllable pair, not necessarily in standard controllable form. Show that there is a 1 by n vector f such that $A - bf$ has arbitrarily prescribed eigenvalues.

The line of reasoning described above can be pushed further to yield the analogue of Exercise 2.1.4 for any number of inputs $m \geq 1$. We shall not discuss this here, as we are following a different path. Instead, we shall construct a stabilizing feedback F using the methods of optimal control. What follows is an outline of the contents of this chapter.

Let (A, B, C) be any triple. Consider (1.1) together with the *cost*

$$J^u(x^0) = \int_0^\infty |u(t)|^2 + |Cx(t)|^2 \, dt \qquad (x(\cdot) = x^u(\cdot)).$$

$J^u(x^0)$ is *the cost corresponding to the control* $u(\cdot)$ *starting at* x^0. Suppose, for the sake of argument, that $m = 1$. We have seen above that a stabilizing feedback f can always be chosen. How much then is the cost incurred in applying this feedback law? More generally, if F is stabilizing and the feedback law (1.2) is implemented, then how much is the corresponding cost?

2.1.5. Proposition. *Suppose that F is a stabilizing feedback law. The cost incurred in applying (1.2) starting from x^0 is*

$$J^F(x^0) = x^{0*}Lx^0,$$

where L is the solution of the Lyapunov equation

$$\bar{A}^*L + L\bar{A} + F^*F + C^*C = 0, \qquad \bar{A} = A - BF.$$

PROOF. Let $x(t) = e^{t\bar{A}}x^0$, $u(t) = -Fx(t)$, $t \geq 0$, be the corresponding state trajectory and control, starting from x^0. Then (Exercise 1.2.11)

$$J^F(x^0) = \int_0^\infty |u(t)|^2 + |Cx(t)|^2 \, dt$$

$$= \int_0^\infty x(t)^*(F^*F + C^*C)x(t) \, dt$$

$$= x^{0*}\left(\int_0^\infty e^{t\bar{A}^*}(F^*F + C^*C)e^{t\bar{A}} \, dt\right)x^0$$

$$= x^{0*}Lx^0,$$

with $Q = F^*F + C^*C$. □

To see how this depends on the choice of F, let us take the simplest case: $n = m = p = 1$, with $b \neq 0$ and $c \neq 0$. Then the solution of the (linear!) Lyapunov equation is

$$l = -\frac{|f^2| + |c|^2}{\bar{a} + \bar{a}^*}, \qquad \bar{a} = a - bf, \qquad \text{Re}(\bar{a}) < 0.$$

Here $a, b, c, f,$ and l are simply complex numbers. Thus *the cost can be made arbitrarily large*, by choosing *the gain f large* (i.e., $bf \to +\infty$). At the same time, this *makes $x(t)$ go to zero faster*, since $x(t) = e^{(a-bf)t}$. The relevant question now is: what feedback f will *minimize the cost*? The answer will provide us not with the fastest way to reach the origin, but with the cheapest, in the sense of least cost. *The analysis of the minimizing feedback law in the general case is the main goal of the chapter.*

A control $u^{\#}(\cdot)$ is *optimal at x^0* if $J^{u^{\#}}(x^0) \leq J^u(x^0)$ for all controls $u(\cdot)$. We shall see that the optimal control is given by feedback F and we shall give an explicit expression for F in terms of A, B, and C.

2.1.6. Exercise. Let $n = m = p = 1$, with $b \neq 0, c \neq 0$. Show that the minimum value of l, as f varies over the set $\{f | \text{Re}(\bar{a}) < 0\}$ is the unique solution $k > 0$ of

$$a^*k + ka = kbb^*k - c^*c,$$

and that the minimizing $f = b^*k$.

2.1.7. Exercise. Let (A, B, C) be any system and let F be an m by p matrix. Compute the transfer function $G_F(s)$ of the triple $(A - BFC, B, C)$ in terms of the transfer function $G(s) = C(sI - A)^{-1}B$. Show that $(A - BFC, B, C)$ is minimal iff (A, B, C) is minimal.

2.1.8. Exercise. Let (A, b, c) be any triple with $g(s) = c(sI - A)^{-1}b$ scalar. Show that

$$\det(sI - A + bc) = \det(sI - A)(1 + c(sI - A)^{-1}b).$$

In particular, conclude that $\det(I + bc) = 1 + cb$ for any vectors b, c. (*Hint*: Do this by first assuming that (A, b, c) is minimal and using 1.4.13, 1.5.5, and 2.1.7, or do this by assuming first that (A, b) is controllable and using 1.4.3.)

2.2. Properness

Consider the system

$$\dot{x} = Ax + Bu, \qquad x(0) = x^0 \text{ in } \mathbb{C}^n, \tag{2.1}$$

$$y = Cx. \tag{2.2}$$

Corresponding to any control $u(\cdot)$ set

$$J^u(x^0) = \int_0^\infty |u(t)|^2 + |y(t)|^2 \, dt;$$

this is the *cost* of the control $u(\cdot)$ *starting from* x^0. If the control arises from feedback (1.2), then instead of $J^u(x^0)$ we write $J^F(x^0)$, where F is the feedback matrix. Set

$$S(x^0) = \min\{J^u(x^0)|\text{all controls } u(\cdot)\}.$$

$S(x^0)$ is the *optimal cost* starting from x^0. Note that either $J^u(x^0) \geq 0$ or $S(x^0) \geq 0$ may equal $+\infty$, and that $S(0) = 0$.

The entire development in this section is based on the following basic fact from analysis.

2.2.1. Lemma. *Let $u_k(\cdot), k \geq 1$, be a sequence of controls with $\int_0^T |u_k(t)|^2 \, dt \leq c_T$ for some constant c_T, for all $T > 0$. Then there is a subsequence $v_k(\cdot), k \geq 1$, and a control $u(\cdot)$ such that*

$$e_k(T; v) = \int_0^T v(t)^*(v_k(t) - u(t)) \, dt \to 0 \qquad \text{as} \quad k \uparrow \infty \qquad (2.3)$$

for all $T > 0$ and all controls $v(\cdot)$.

Using 2.2.1, we derive a basic continuity result for $J^u(x^0)$.

2.2.2. Corollary. *Let $u_k(\cdot)$ be any sequence of controls, suppose that $x_k^0 \to x^0$ and suppose that $J^{u_k}(x_k^0)$ converges to some limit, as $k \uparrow \infty$. Then there is a control $u(\cdot)$ satisfying*

$$J^u(x^0) \leq \lim_{k \to \infty} J^{u_k}(x_k^0). \qquad (2.4)$$

In particular, for each x^0 there is a $u^\#(\cdot)$ with $S(x^0) = J^{u^\#}(x^0)$, and S is a lower semicontinuous function: with $x_k^0 \to x^0$ as $k \uparrow \infty$, one has

$$S(x^0) \leq \lim_{k \uparrow \infty} S(x_k^0) \qquad (2.5)$$

whenever the limit on the right-hand side exists.

PROOF. By 2.2.1 there is a subsequence $v_k(\cdot)$ and a control $u(\cdot)$ such that (2.3) holds. Let $x_k(t)$ denote $x^{v_k}(t; x_k^0), k \geq 1$. Let $w(\cdot)$ be an $\mathbb{R}^n$-valued function of time and set $v(t) = \int_t^T B^* e^{(s-t)A^*} w(s) \, ds, t \geq 0$. Then

$$e_k'(T; w) = \int_0^T w(t)^*(x_k(t) - x^u(t)) \, dt$$

$$= e_k(T; v) + \left(\int_0^T w(t)^* e^{At} \, dt\right)(x_k^0 - x^0)$$

$$\to 0 \qquad \text{as} \quad k \uparrow \infty.$$

Let $J_T^u(x^0) = \int_0^T |u(t)|^2 + |y(t)|^2 \, dt$ be the cost up to time T; then by the linearity of (2.1) and (2.2) and the quadratic nature of J_T,

$$0 \leq J_T^{v_k - u}(x_k^0 - x^0) = J_T^{v_k}(x_k^0) - J_T^u(x^0) - 2e_k(T; u) - 2e'_k(T; x^u)$$

for all $T > 0$ and all $k \geq 1$. If we then let $k \uparrow \infty$ and $T \uparrow \infty$, in that order, in this last equation, (2.4) follows. To prove the second statement, for each $k \geq 1$ choose $u_k(\cdot)$ such that $J^{u_k}(x^0) \leq S(x^0) + 1/k$ and apply (2.4). The last statement follows by choosing $u_k(\cdot)$ such that $J^{u_k}(x_k^0) = S(x_k^0)$ for all $k \geq 1$. □

We say that a control $u(\cdot)$ is *finite* at x^0 if $J^u(x^0)$ is finite. Note that $S(x^0)$ is finite iff there is at least one control $u(\cdot)$ finite at x^0.

2.2.3. Exercise. Let $n = m = p = 1$, and let $a = b = 0$, $c = 1$; graph S.

2.2.4. Lemma. *Let* (A, B) *be stabilizable. Then for all* x^0, $S(x^0)$ *is finite.*

PROOF. Let F be a stabilizing feedback with $\bar{A} = A - BF$. Then, according to 2.1.5, $J^F(x^0) = x^{0*}Lx^0 < +\infty$. Thus $S(x^0) < +\infty$. □

2.2.5. Exercise. Let x^0 be controllable to zero. Show that $S(x^0)$ is finite. Conclude that (A, B) controllable implies $S(x^0)$ is finite for all x^0.

2.2.6. Exercise. Show that $S(x^0) \neq 0$ iff x^0 is observable from zero. Conclude that (A, C) is observable iff $S(x^0) \neq 0$ for all $x^0 \neq 0$.

Thus when (A, B, C) is minimal, $0 < S(x^0) < +\infty$ whenever $x^0 \neq 0$. We now define a crucial concept: a function $\varphi \colon \mathbb{C}^n \to [0, +\infty]$ is *proper* if for any sequence of states $x_1, x_2, \ldots$ tending to ∞, $|x_k| \to +\infty$, one has $\varphi(x_k) \to +\infty$. This is the same as saying that for all finite M, the set $\{x \mid \varphi(x) \leq M\}$ is a bounded subset of $\mathbb{C}^n$.

2.2.7. Exercise. Let S be as in 2.2.3. Is S proper? Which of $\varphi(x) = \sin(x)$, $\varphi(x) = |x|^2$, $\varphi(x) = e^{x^2}$ are proper?

We say that (A, B, C) is *proper* if S is proper. The importance of properness is brought out in the following theorem.

2.2.8. Theorem. *Let* (A, C) *be observable and let* (A, B, C) *be proper. If* $u(\cdot)$ *is finite at* x^0, *then* $u(\cdot)$ *is stabilizing at* x^0.

PROOF. We are given that $J^u(x^0) < +\infty$. Set $v(t) = u(t + T)$, $t \geq 0$, and $w = u *_T v$. Then (Exercise 1.3.3) $w = u$ and

$$y^v(t; x^u(T)) = y^w(t + T) = y^u(t + T), \qquad t \geq 0.$$

Now

$$S(x^u(T)) \leq J^v(x^u(T))$$

$$= \int_0^\infty |v(t)|^2 + |y^v(t; x^u(T))|^2 \, dt$$

$$= \int_0^\infty |u(t + T)|^2 + |y^u(t + T)|^2 \, dt$$

$$= \int_T^\infty |u(t)|^2 + |y(t)|^2 \, dt \tag{2.6}$$

Thus $S(x^u(T)) \to 0$ as $T \uparrow \infty$. Hence, by properness of S, $x^u(T)$, $T \geq 0$, is forced to remain in a bounded set in $\mathbb{C}^n$. Now if $0 < T_1 < T_2 < \cdots \uparrow \infty$ is any sequence of times with $x^u(T_k) \to x$ as $k \uparrow \infty$, then by lower semicontinuity of S, $\lim_{k \uparrow \infty} S(x^u(T_k)) \geq S(x)$. This implies that $S(x) = 0$. Observability now implies (Exercise 2.2.6) that $x = 0$. Thus the only limiting state of $x^u(T)$, as $T \uparrow \infty$, is the origin. This shows that $x^u(T) \to 0$ as $T \uparrow \infty$. $\quad\square$

We now show that as a consequence of linearity (this is not true in general) a triple (A, B, C) is proper iff (A, C) is observable! First we need a lemma.

2.2.9. Lemma. S *is quadratic (homogeneous of degree* 2*):* $S(\lambda x^0) = |\lambda|^2 S(x^0)$ *for any complex number* λ.

PROOF. If $\lambda = 0$ we already know this. So assume that $\lambda \neq 0$. By 1.3.2, $x^{\lambda u}(t; \lambda x^0) = \lambda x^u(t; x^0)$, $t \geq 0$. Thus $|\lambda u|^2 + |y^{\lambda u}|^2 = |\lambda|^2(|u|^2 + |y^u|^2)$. Hence $J^{\lambda u}(\lambda x^0) = |\lambda|^2 J^u(x^0)$. Now $u(\cdot) \to \lambda u(\cdot) = v(\cdot)$ is a permutation of the set of all controls. Thus

$$S(\lambda x^0) = \min\{J^v(\lambda x^0)|\text{all controls } v(\cdot)\}$$

$$= \min\{J^{\lambda u}(\lambda x^0)|\text{all controls } u(\cdot)\}$$

$$= \min\{|\lambda|^2 J^u(x^0)|\text{all controls } u(\cdot)\}$$

$$= |\lambda|^2 S(x^0). \quad\square$$

2.2.10. Proposition. *Observability of* (A, C) *is equivalent to properness of* S.

PROOF. If (A, C) is not observable, then there is an $x^0 \neq 0$ with $S(x^0) = 0$. Let $x_k = k x^0$. Then $|x_k| \to \infty$ as $k \uparrow \infty$ but $S(x_k) = S(k x^0) = k^2 S(x^0) = 0$ for all k. Thus S is not proper. On the other hand, if (A, C) is observable, then $S(x^0) > 0$ for $|x^0| = 1$. Because S is lower semicontinuous, we then have

$$\min\{S(x)||x| = 1\} > 0.$$

(This need not be true without lower semicontinuity.) Let x_k be any sequence with $|x_k| \to \infty$. Set $\lambda_k = |x_k|$. Then

$$S(x_k) = S\left(\frac{\lambda_k x_k}{\lambda_k}\right) = \lambda_k^2 S\left(\frac{x_k}{\lambda_k}\right)$$

$$\geq \lambda_k^2 \min\{S(x)|\,|x| = 1\} \to +\infty \qquad \text{as} \quad k\uparrow\infty.$$

Thus S is proper.   □

2.2.11. Corollary. *Let* (A, C) *be observable. If* $u(\cdot)$ *is finite at* x^0, *then* $u(\cdot)$ *is stabilizing at* x^0.

PROOF. Combine 2.2.8 and 2.2.10.   □

We stated 2.2.8 separately because it is the result that continues to hold in the nonlinear situation; 2.2.11 is not true in general.

2.3. Optimal Control

Motivated by Exercise 2.1.6, consider the following equation

$$A^*K + KA = KBB^*K - C^*C. \tag{3.1}$$

2.3.1. Exercise. Given (A, B, C), let $\Phi = \{F|\bar{A} = A - BF$ is stable$\}$. Corresponding to each such F, let L be the solution of the Lyapunov equation $(L = L(F)$, see 2.1.5)

$$\bar{A}^*L + L\bar{A} + F^*F + C^*C = 0. \tag{3.2}$$

Then L is self-adjoint. Given two much matrices L_1 and L_2, recall that $L_1 \geq L_2$ if $x^*L_1 x \geq x^*L_2 x$ for all x in $\mathbb{C}^n$.

Suppose there is an F_0 in Φ such that $K = L(F_0)$ satisfies $K \leq L(F)$ for all F in Φ. Show that K then satisfies (3.1), and that the feedback F_0 that gave rise to K satisfies $F_0 = B^*K$. (*Hint:* Plug in (3.2) $F_\varepsilon = F_0 + \varepsilon D$ with D arbitrary, and let L_ε be the solution, and take derivatives with respect to ε.)

2.3.2. Exercise. If (3.1) has a unique solution K, then show $K = K^*$.

2.3.3. Exercise. Suppose that $K = K^*$ satisfies (3.1) with $\bar{A} = A - BB^*K$ stable. Show that $K \geq 0$. Also if $(\bar{A}, C)$ is observable, show that $K > 0$.

For $T > 0$ set

$$J_T^u(x^0) = \int_0^T |u(t)|^2 + |y(t)|^2 \, dt.$$

Equation (3.1) is the *Algebraic Riccati Equation* (ARE).

2.3.4. Lemma. *Suppose that $K = K^*$ is a solution of* (3.1). *Then*

$$J_T^u(x^0) = -x(T)^*Kx(T) + x^{0*}Kx^0 + \int_0^T |u(t) + B^*Kx(t)|^2 \, dt, \quad (3.3)$$

with $x(\cdot) = x^u(\cdot)$.

PROOF.

$$x(T)^*Kx(T) - x^{0*}Kx^0 = \int_0^T \frac{d}{dt}(x(t)^*Kx(t)) \, dt$$

$$= \int_0^T \dot{x}^*Kx + x^*K\dot{x} \, dt$$

$$= \int_0^T (Ax + Bu)^*Kx + x^*K(Ax + Bu) \, dt$$

$$= \int_0^T x^*(A^*K + KA)x + 2u^*B^*Kx \, dt$$

$$= \int_0^T x^*(KBB^*K - C^*C)x + 2u^*B^*Kx \, dt$$

$$= \int_0^T |B^*Kx|^2 - |Cx|^2 - |u|^2 + |u|^2 + 2u^*B^*Kx \, dt$$

$$= \int_0^T -|u|^2 - |y|^2 + |u + B^*Kx|^2 \, dt$$

$$= \int_0^T |u(t) + B^*Kx(t)|^2 \, dt - J_T^u(x^0). \qquad \square$$

2.3.5. Lemma. *Let (A, C) be observable. Let $K = K^*$ satisfy equation* (3.1). *Then for all $u(\cdot)$ finite at x^0,*

$$J^u(x^0) = x^{0*}Kx^0 + \int_0^\infty |u(t) + B^*Kx(t)|^2 \, dt. \quad (3.4)$$

PROOF. By 2.2.11, $u(\cdot)$ is stabilizing at x^0, i.e., $x(T) \to 0$ as $T \uparrow \infty$. The result follows by letting $T \uparrow \infty$ in (3.3). $\qquad \square$

2.3.6. Lemma. *Let (A, C) be observable. Then a solution $K = K^*$ of the Riccati equation is nonnegative, $K \geq 0$, iff $\bar{A} = A - BB^*K$ is stable.*

PROOF. If $\bar{A}$ is stable, Exercise 2.3.3 shows that $K \geq 0$. Suppose now that $K \geq 0$ and consider the feedback law $u = -Fx = -B^*Kx$. Equation (3.3) that implies that $J_T^F(x^0) = -x(T)^*Kx(T) + x^{0*}Kx^0 \leq x^{0*}Kx^0 < +\infty$. Letting $T \uparrow \infty$, we see that $J^F(x^0)$ is finite and so F is stabilizing. Thus $\bar{A}$ is stable. $\qquad \square$

The following is the main result of the chapter.

2.3.7. Theorem. *Let (A, C) be observable and let $K \geq 0$ be a solution of the algebraic Riccati equation. Then:*

(i) $S(x^0) = x^{0*}Kx^0$ *for all* x^0.
(ii) K *is the* only *nonnegative solution of* (3.1).
(iii) K *is positive:* $K > 0$.
(iv) *The feedback* $F = B^*K$ *is stabilizing, i.e.,* $\bar{A} = A - BB^*K$ *is stable.*
(v) *For each* x^0, *there is a unique optimal control* $u^{\#}(\cdot)$. *This control is given by feedback* $F = B^*K$.

PROOF. Fix x^0 in $\mathbb{C}^n$. Equation (3.4) shows that $J^u(x^0) \geq x^{0*}Kx^0$ for all $u(\cdot)$ finite at x^0 and hence all $u(\cdot)$. Thus $S(x^0) \geq x^{0*}Kx^0$. Let $u^{\#}(\cdot)$ be given by feedback $F = B^*K$. Then, by 2.3.4, $J^{u^{\#}}(x^0) \leq x^{0*}Kx^0$ and so $u^{\#}(\cdot)$ is optimal. This verifies (i). Now if $u(\cdot)$ is any other optimal control, then by (3.4) $u(\cdot)$ satisfies the feedback law $u = -Fx$; by 2.1.1, $u(\cdot)$ must equal $u^{\#}(\cdot)$. This verifies (v). Now since S is uniquely determined, (i) implies (ii). Also since x^0 is arbitrary, (v) implies (iv). Finally, (i) and 2.2.6 imply (iii). $\qquad\square$

Theorem 2.3.7 almost completely answers all our questions concerning the optimal control of $J^u(x^0)$. There is one nagging detail, however: we do not know whether (3.1) has any solutions K at all! *It is for this point that we need controllability.* In the next section we shall see that controllability implies the existence of at least one *nonnegative* solution K to (3.1). Coupled with Theorem 2.3.7, we see that minimality guarantees the existence of a unique solution to (3.1). We emphasize that the existence of a matrix K is not enough; K has to be *nonnegative*. In Chapter 5 we shall see an example of an analogous situation where a solution to a certain equation exists but is not nonnegative, and where the result corresponding to 2.3.7(iv) fails (see 5.5.6).

As a corollary of the above we have the following useful estimate.

2.3.8. Corollary. *Let (A, C) be observable and let $K \geq 0$ be a solution of the algebraic Riccati equation. Then for any $u(\cdot)$ and $T \geq 0$*

$$x^u(T)^*Kx^u(T) \leq \int_T^\infty |u(t)|^2 + |Cx^u(t)|^2 \, dt.$$

PROOF. This is just (2.6). $\qquad\square$

Let $K = K^*$ be a solution of the algebraic Riccati equation. We list some associated transfer functions. If $G(s) = C(sI - A)^{-1}B$, then set

$$\bar{G}(s) = C(sI - \bar{A})^{-1}B, \qquad \bar{A} = A - BB^*K,$$

$$G^{\#}(s) = B^*K(sI - A)^{-1}B,$$

$$\bar{G}^{\#}(s) = B^*K(sI - \bar{A})^{-1}B, \qquad \bar{A} = A - BB^*K.$$

These transfer functions depend only on $G(s)$ and not on the particular realization (A, B, C) (see 2.5.5).

2.3.9. Exercise. Show that

$$I + G(-s^*)^*G(s) = (I + G^\#(-s^*))^*(I + G^\#(s)).$$

2.3.10. Exercise. Show that

$$I - \bar{G}(-s^*)^*\bar{G}(s) = (I - \bar{G}^\#(-s^*))^*(I - \bar{G}^\#(s)).$$

2.3.11. Exercise. Show that

$$(I - \bar{G}^\#(s))^{-1} = I + G^\#(s).$$

Define $(\tilde{A}, \tilde{B}, \tilde{C})$ by setting

$$\tilde{A} = \begin{pmatrix} A & BB^* \\ C^*C & -A^* \end{pmatrix}, \qquad \tilde{B} = \begin{pmatrix} B \\ 0 \end{pmatrix}, \qquad \tilde{C} = (0, -B^*).$$

The system $(\tilde{A}, \tilde{B}, \tilde{C})$ then has dimension $2n$, m inputs, and m outputs. Set

$$\tilde{G}(s) = \tilde{C}(sI - \tilde{A})^{-1}\tilde{B}.$$

2.3.12. Exercise. Compute $\tilde{G}(s)$ in terms of $G(s)$. (*Hint*: Compute $\tilde{C}(\tilde{A} + \tilde{B}\tilde{C})^k\tilde{B}$ first, then use 2.1.7.)

Fix a solution $K = K^*$ of the algebraic Riccati equation. Let V be the n-dimensional subspace of $\mathbb{C}^{2n}$ of all vectors of the form

$$\tilde{x} = \begin{pmatrix} x \\ -Kx \end{pmatrix}.$$

Set $\bar{A} = A - BB^*K$.

2.3.13. Exercise. Show that $(\tilde{A}, \tilde{B}, \tilde{C})$ is Hamiltonian (1.5.4) and that $\tilde{A}\tilde{x} = (\bar{A}x)^\sim$. Conclude that $\det(sI - \tilde{A}) = \det(sI - \bar{A})\det(sI + \bar{A}^*)$.

Suppose that there exists a nonnegative solution $K \geq 0$ of the algebraic Riccati equation. Then, according to 2.3.6, $\bar{A}$ is stable; in this case we have that $\tilde{A}$ is *hyperbolic*, i.e., the eigenvalues of $\tilde{A}$ lie off the imaginary axis.

2.3.14. Exercise. Suppose that $(\tilde{A}, \tilde{B}, \tilde{C})$ is minimal. Show that (A, B, C) is then minimal, and that (A, B, B^*K) is also minimal.

The next exercise shows that the converse to 2.3.14 is false.

2.3.15. Exercise. Let $p(s) = s + 1$, $q(s) = s^2 - s$. Set $g(s) = p(s)/q(s)$. Compute the transfer functions $\bar{g}(s)$, $g^\#(s)$, $\bar{g}^\#(s)$, and $\tilde{g}(s)$.

2.4. The Riccati Equation

The purpose of this section is to establish the existence of at least one solution $K \geq 0$ to the Riccati equation when (A, B) is controllable or more generally stabilizable.

We begin with a preliminary lemma.

2.4.1. Lemma. *Let $A(t), 0 \leq t \leq T$, be an n by n matrix-valued function of time such that*

$$a = \int_0^T |A(t)| \, dt$$

is finite. Then for each x^0 in $\mathbb{C}^n$, there is a unique solution $x(t), 0 \leq t \leq T$, in $\mathbb{C}^n$ of

$$\dot{x} = A(t)x, \qquad x(0) = x^0. \tag{4.1}$$

PROOF. Define $A_0(t) = I, 0 \leq t \leq T$, and for $k = 1, 2, \ldots$

$$A_k(t) = \int_0^t A(s)A_{k-1}(s) \, ds, \qquad 0 \leq t \leq T.$$

Then we claim that $|A_k(t)| \leq (1/k!)a(t)^k$, where $a(t) = \int_0^t |A(s)| \, ds$. Indeed, assume this is so for $k = N$; then

$$|A_{N+1}(t)| \leq \int_0^t |A(s)| \, |A_N(s)| \, ds$$

$$\leq \int_0^t \dot{a}(s) \frac{1}{N!} a(s)^N \, ds$$

$$= \frac{1}{(N+1)!} a(t)^{N+1}.$$

This verifies the claim; now set

$$x(t) = x^0 + A_1(t)x^0 + A_2(t)x^0 + \cdots .$$

Because of the bound on $|A_k(t)|$, this series converges uniformly for $0 \leq t \leq T$. Hence $x(\cdot)$ is a continuous function of time. Now

$$\int_0^t A(s)x(s) \, ds = \int_0^t \sum_{k=0}^\infty A(s)A_k(s)x^0 \, ds$$

$$= \sum_{k=0}^\infty \left(\int_0^t A(s)A_k(s) \, ds \right) x^0$$

$$= \sum_{k=0}^\infty A_{k+1}(t)x^0$$

$$= x(t) - x^0.$$

Thus $x(\cdot)$ satisfies (4.1).   □

2.4.2. Exercise. Show that (4.1) has at most one solution.

Consider now the following differential equation

$$\dot{M} = A^*M + MA - MBB^*M + C^*C, \qquad t \geq 0, \quad M(0) = 0. \quad (4.2)$$

2.4.3. Lemma. *If* (4.2) *has a solution* $M(t)$, $0 \leq t \leq T$, *then*

$$x^{0*}M(T)x^0 = \min\{J_T^u(x^0)|all\ controls\ u(\cdot)\}, \quad (4.3)$$

and, moreover, $M(t) = M(t)^*$, $M(t) \geq 0$, $0 \leq t \leq T$.

PROOF. Let $x = x^u(t)$ and $M = M(T - t)$, $0 \leq t \leq T$. Then

$$x^{0*}M(T)x^0 = x^{0*}M(T)x^0 - x(T)^*M(0)x(T)$$

$$= \int_0^T -\frac{d}{dt}x(t)^*M(T - t)x(t)\ dt$$

$$= \int_0^T -\dot{x}^*Mx + x^*\dot{M}x - x^*M\dot{x}\ dt$$

$$= \int_0^T -(Ax + Bu)^*Mx + x^*\dot{M}x - x^*M(Ax + Bu)\ dt$$

$$= \int_0^T x^*(-A^*M + \dot{M} - MA)x - 2u^*B^*Mx\ dt \quad (4.4)$$

$$= \int_0^T |u|^2 + |y|^2 - |u + B^*Mx|^2\ dt$$

$$= J_T^u(x^0) - \int_0^T |u(t) + B^*M(T - t)x^u(t)|^2\ dt$$

$$\leq J_T^u(x^0).$$

Thus $x^{0*}M(T)x^0 \leq \min\{J_T^u(x^0)|all\ controls\ u(\cdot)\}$. On the other hand, consider the linear time-varying equation

$$\dot{x} = A(t)x = (A - BB^*M(T - t))x, \qquad 0 \leq t \leq T, \quad x(0) = x^0.$$

Let $x^\#(t)$, $0 \leq t \leq T$, be the solution whose existence is guaranteed by 2.4.1. Set $u^\#(t) = -B^*M(T - t)x^\#(t)$, $0 \leq t \leq T$. Then $x^\#(\cdot) = x^{u^\#}(\cdot)$ and so by (4.4) $J_T^{u^\#}(x^0) = x^{0*}M(T)x^0$. This shows (4.3). Note that $M(t)^*$, $0 \leq t \leq T$, also satisfies (4.2). By uniqueness of solutions of a differential equation, we must have $M(t) = M(t)^*$, $0 \leq t \leq T$. In deriving (4.4) above, we used the self-adjointness of M. Moreover, (4.3) shows that $M(T)$ is nonnegative, $M(T) \geq 0$. □

2.4.4. Corollary. *Equation* (4.2) *has a unique solution* $M(t)$ *defined for all time* $t \geq 0$. *For each* x^0, $x^{0*}M(t)x^0$ *is an increasing function of* $t \geq 0$, *and*

$$x^{0*}M(t)x^0 \leq S(x^0), \qquad t \geq 0. \quad (4.5)$$

PROOF. Since (4.2) is an ordinary differential equation in the n^2 entries of M, it has a unique solution on some interval of time $[0, \tau)$. Let $\tau \leq +\infty$ be the largest number for which this is so. Now (4.3) shows that $x^{0*}M(t)x^0$ is an increasing function of t, since $J_t^u(x^0)$ is so for each $u(\cdot)$. If τ is finite, then by (4.3)

$$\lim_{T \uparrow \tau} x^{0*}M(T)x^0 \leq J_\tau^0(x^0) < +\infty$$

exists for each x^0. Since $M(t)$ is self-adjoint, this shows that the limit of each entry of $M(t)$ exists as $t \uparrow \tau$. Now by solving (4.2) starting from time τ, we can extend the solution $M(t)$ to be defined after time τ. This contradicts the maximality of τ, and hence $\tau = +\infty$. Now since $J_T^u(x^0) \leq J^u(x^0)$ (4.5) follows from (4.3). □

We show now that the finiteness of $S(x^0)$ for all x^0 is enough to guarantee the existence of K.

2.4.5. Proposition. *Suppose that $S(x^0)$ is finite for all x^0. Then there is a nonnegative solution $K \geq 0$ to the algebraic Riccati equation (3.1).*

PROOF. Let $T \uparrow \infty$ in (4.5). Since $S(x^0)$ is finite, the limit exists for each x^0. Since $M(T)$ is self-adjoint, the limit of $cM(T)b$ as $T \uparrow \infty$ must exist for each b and c. This limit, call it $Q(b, c)$, is linear in b and linear in c. Therefore this limit must be of the form cKb for some matrix K. Thus $\lim_{T \uparrow \infty} M(T) = K$ exists. Note that because $M(T)$ is nonnegative, so is K. Now

$$A^*K + KA = \int_0^1 A^*K + KA \, dt$$

$$= \int_0^1 \lim_{T \uparrow \infty} A^*M(t + T) + M(t + T)A \, dt$$

$$= \int_0^1 \lim_{T \uparrow \infty} M(t + T)BB^*M(t + T) - C^*C + \dot{M}(t + T) \, dt$$

(by (4.2))

$$= \int_0^1 KBB^*K - C^*C + \lim_{T \uparrow \infty} \dot{M}(t + T) \, dt$$

$$= KBB^*K - C^*C + \lim_{T \uparrow \infty} \int_0^1 \dot{M}(t + T) \, dt$$

$$= KBB^*K - C^*C + \lim_{T \uparrow \infty} (M(T + 1) - M(T))$$

$$= KBB^*K - C^*C + (K - K) = KBB^*K - C^*C. \qquad \square$$

2.4.6. Corollary. *If (A, B) is stabilizable, or if (A, B) is controllable, then there exists at least one nonnegative solution K to (3.1).*

PROOF. Follows from 2.2.4 and 2.2.5. □

2.4.7. Corollary. *If (A, B) is controllable, then (A, B) is stabilizable.*

PROOF. By 2.4.6 there is a nonnegative solution K to (3.1). By 2.3.7 the feedback $F = B^*K$ is stabilizing. □

Combining the results of the last section with the above, we see that to each minimal (A, B, C), there is a unique positive solution of the algebraic Riccati equation. In the next section we look at some of its properties.

2.4.8. Exercise. Use (4.3) to conclude that there is a $c = c(A, B, C) > 0$ such that for all $u(\cdot)$

$$\int_0^\infty |x^u(t)|^2 \, dt \le c \int_0^\infty |u(t)|^2 + |Cx^u(t)|^2 \, dt.$$

2.5. The Space $M(m, n, p)$

Fix the state dimension n, the number of inputs m, and the number of outputs p. The set of all triples (A, B, C) can then be put into one-to-one correspondence with the vector space $\mathbb{C}^N$, where $N = n^2 + nm + np$. Let $\tilde{M}(m, n, p)$ denote the subset consisting of all minimal such triples. Then $\tilde{M}(m, n, p)$ is an *open* subset of $\mathbb{C}^N$ in the sense that any triple (A', B', C') sufficiently close to a minimal triple (A, B, C) is also minimal. By "close" we mean that the corresponding entries of the matrices are close.

2.5.1. Exercise. Verify this.

Corresponding to any (A, B, C) in $\tilde{M}(m, n, p)$, we let $K(A, B, C)$ denote the unique nonnegative solution $K > 0$ of the algebraic Riccati equation (3.1). This section will deal with some aspects of the dependence of $K = K(A, B, C)$ on (A, B, C). Note that the complement of $\tilde{M}(m, n, p)$,

$$\mathbb{C}^N - \tilde{M}(m, n, p) \subset \mathbb{C}^N,$$

is described as exactly *the set of all triples for which certain polynomials in the entries of A, B, C vanish.* Thus $\mathbb{C}^N - \tilde{M}(m, n, p)$ is what is known as an "algebraic set."

2.5.2. Exercise. Which polynomials?

We now let $M(m, n, p)$ be the set of all p by m transfer functions $G(s)$ which are realizable by a minimal triple (A, B, C) of dimension n. Let GL_n denote the set of all invertible n by n matrices P. GL_n is a "group" in the following sense.

2.5.3. Exercise. Show that if P and Q are in GL_n, then so is PQ, P^{-1}, and I.

In Chapter 1 we studied the map $\pi: \tilde{M}(m, n, p) \to M(m, n, p)$ given by

$$\pi(A, B, C) = C(sI - A)^{-1}B = G(s).$$

This map is onto, by definition, but not one-to-one. Indeed (A_0, B_0, C_0) and (A, B, C) have the same transfer function iff there is a *unique* P in GL_n such that $PA_0P^{-1} = P$, $PB_0 = B$, and $C_0P^{-1} = C$ (1.5.3). Thus the inverse image of a given transfer function under the map π is in one-to-one correspondence with GL_n. Since there are n^2 parameters in GL_n, it seems reasonable that the map π "erases" n^2 *parameters from* $\tilde{M}(m, n, p)$ and so the set $M(m, n, p)$ is describable by $N - n^2 = nm + np$ parameters. It turns out that this is the case. For example, for single-input–single-output systems

$$g(s) = \frac{p_n s^{n-1} + \cdots + p_2 s + p_1}{s^n + q_n s^{n-1} + \cdots + q_1} = \frac{p(s)}{q(s)}$$

with $p(s)$ and $q(s)$ having no common factors; thus a parametrization of $M(1, n, 1)$ may be given by

$$g(s) \to (p_1, \ldots, p_n, q_1, \ldots, q_n)$$

explicitly exhibiting $M(1, n, 1)$ as an open subset of $\mathbb{C}^{2n}$.

Corresponding to each (A, B, C) in $\tilde{M}(m, n, p)$ let $K = K(A, B, C)$ denote the positive solution of (3.1).

2.5.4. Proposition. *The map* $(A, B, C) \to K$ *is a differentiable map of class* C^∞.

PROOF. Write the algebraic Riccati equation (3.1) as $f(A, B, C; K) = 0$ thinking of f as a polynomial map from $\tilde{M}(m, n, p) \times \{$self-adjoint n by n matrices$\}$ to $\{$self-adjoint n by n matrices$\}$. We will use the implicit function theorem. To this end we have to compute

$$D_K f(A, B, C; K)L = \left.\frac{d}{d\varepsilon}\right|_{\varepsilon=0} f(A, B, C; K_\varepsilon) \qquad (K_\varepsilon = K + \varepsilon L)$$

$$= \left.\frac{d}{d\varepsilon}\right|_{\varepsilon=0} A^*K_\varepsilon + K_\varepsilon A - K_\varepsilon BB^*K_\varepsilon + C^*C$$

$$= A^*L + LA - KBB^*L - LBB^*K$$

$$= \bar{A}^*L + L\bar{A} \qquad (\bar{A} = A - BB^*K).$$

Let Q be any n by n self-adjoint matrix. Then since $\bar{A}$ is stable, the Lyapunov equation

$$D_K f(A, B, C; K)L = Q$$

can always be solved for $L = L^*$. Thus the partial derivative of f with respect to K is onto (has full rank) and hence, by the implicit function theorem, the result follows. $\qquad\square$

Actually the above proof shows that K is a real-analytic function of (A, B, C), since the map f is such. Note however that K is not a complex-analytic function of (A, B, C). Consider now the maps

$$(A, B, C) \rightarrow (PAP^{-1}, PB, CP^{-1}) \qquad (P \text{ in } GL_n) \qquad (5.1)$$

and

$$(A, B, C) \rightarrow (-A^*, -C^*, B^*) \qquad (5.2)$$

The map (5.1) goes from $\tilde{M}(m, n, p)$ to $\tilde{M}(m, n, p)$ while the map (5.2) goes from $\tilde{M}(m, n, p)$ to $\tilde{M}(p, n, m)$. Note that as we know the transfer function is unchanged under the map (5.1), while under the map (5.2) $G(s)$ becomes $G(-s^*)^*$. Note also that (5.2) is an involution, i.e., (5.2) applied twice yields the original triple.

2.5.5. Proposition.

$$K(A, B, C) = P^*K(PAP^{-1}, PB, CP^{-1})P, \qquad (5.3)$$

$$K(-A^*, \pm C^*, \pm B^*) = K(A, B, C)^{-1}. \qquad (5.4)$$

PROOF. To prove (5.3) note that the I/O map corresponding to (A, B, C) and starting at x^0 equals the I/O map corresponding to (PAP^{-1}, PB, CP^{-1}) and starting at Px^0. Thus the corresponding costs must be equal for any control $u(\cdot)$, and hence the corresponding optimal costs must also be equal:

$$x^{0*}K(A, B, C)x^0 = (Px^0)^*K(PAP^{-1}, PB, CP^{-1})(Px^0).$$

This verifies (5.3). For (5.4), pre- and post-multiply (3.1) by K^{-1} and $-K^{-1}$, respectively. This yields

$$-AK^{-1} - K^{-1}A^* = -BB^* + K^{-1}C^*CK^{-1}.$$

But this is the Riccati equation for $(-A^*, \pm C^*, \pm B^*)$. $\square$

2.5.6. Exercise. Show that for (A, B, C) Hamiltonian (1.5.4) one has $JK = K^{-1}J$; for (A, B, C) skew, $A^* = -A$, $B^* = C$, one has $K^2 = I$.

Let $\bar{G}(s)$ denote the transfer function of the optimally controlled system,

$$\bar{G}(s) = C(sI - \bar{A})^{-1}B, \qquad \bar{A} = A - BB^*K. \qquad (5.5)$$

By (5.3), $\bar{G}(s)$ depends only on $G(s)$ and not on the particular realization (A, B, C).

On the other hand, let $\bar{G}(s)$ be an arbitrary p by m transfer function. We say that $\bar{G}(s)$ is *bounded real* if

(i) $\bar{G}(s)$ is stable, and
(ii) $I - \bar{G}(i\omega)^*\bar{G}(i\omega) \geq 0$ for all real ω.

2.5.7. Proposition. *Let (A, B, C) be minimal, let $K = K(A, B, C)$, $\bar{A} = A - BB^*K$, and let $\bar{G}(s) = C(sI - \bar{A})^{-1}B$. Then $\bar{G}(s)$ is bounded real.*

PROOF. Any pole of $\bar{G}(s)$ is necessarily a pole of $\bar{A}$. Thus $\bar{G}(s)$ is stable. Also set $\bar{G}^\#(s) = B^*K(sI - \bar{A})^{-1}B$. Then by 2.3.10

$$I - \bar{G}(i\omega)^*\bar{G}(i\omega) = (I - \bar{G}^\#(i\omega))^*(I - \bar{G}^\#(i\omega)) \geq 0. \qquad \square$$

2.5.8. Theorem. *Let $\bar{G}(s)$ be bounded real. Then there is a minimal triple (A, B, C) with $\bar{G}(s) = C(sI - \bar{A})^{-1}B$ and $\bar{A} = A - BB^*K$, where $K = K(A, B, C)$. Moreover, A can be chosen so that all its eigenvalues λ satisfy $\operatorname{Re}(\lambda) \leq 0$.*

The proof of this theorem is complicated; we shall break it up into a series of lemmas and exercises.

Let $(\bar{A}, B, C)$ be a minimal realization of a given bounded real transfer function $\bar{G}(s)$. Then $\bar{A}$ is stable.

2.5.9. Exercise. Let $0 \leq \alpha < 1$. Suppose that $K = K(\alpha) \geq 0$ satisfies

$$\bar{A}^*K + K\bar{A} + KBB^*K + \alpha C^*C = 0. \qquad (5.6)$$

Set $A = A(\alpha) = \bar{A} + BB^*K(\alpha)$,

$$G_\alpha^\#(s) = B^*K(sI - A)^{-1}B, \qquad (5.7)$$

$$\bar{G}_\alpha^\#(s) = B^*K(sI - \bar{A})^{-1}B. \qquad (5.8)$$

Show that for all ω real,

$$(I - \bar{G}_\alpha^\#(i\omega))^*(I - \bar{G}_\alpha^\#(i\omega)) \geq (1 - \alpha)I.$$

(*Hint:* See 2.5.7.)

2.5.10. Exercise. Referring to 2.5.9, show that for $0 < \alpha < 1$, $K = K(\alpha) > 0$.

2.5.11. Exercise. Let $0 < \alpha < 1$. Let λ be an eigenvalue of A, $Ax = \lambda x$ for some $x \neq 0$. Show that *either* λ is a pole of $G_\alpha^\#(s)$ *or* λ is stable, $\operatorname{Re}(\lambda) < 0$. (*Hint:* See 1.5.6.)

2.5.12. Exercise. Show that

$$(I - \bar{G}_\alpha^\#(s))^{-1} = I + G_\alpha^\#(s);$$

conclude that

$$|I + G_\alpha^\#(i\omega)|^2 \leq \frac{n}{1 - \alpha} \qquad \text{for all real } \omega.$$

2.5.13. Lemma. *Let $0 < \alpha < 1$ and let $K \geq 0$ solve (5.6). Then the eigenvalues of $A = \bar{A} + BB^*K$ are not on the imaginary axis.*

PROOF. Let $Ax = \lambda x$, with $x \neq 0$. By 2.5.11, either λ is a pole of $G_\alpha^\#(s)$ or we have nothing to prove. So assume that λ is a pole of $G_\alpha^\#(s)$. By 2.5.12, this means that λ is a zero of $I - \bar{G}_\alpha^\#(s)$. Thus, by 2.5.9, λ is not on the imaginary axis. □

Now let $\Phi = \{K = K^* | A = \bar{A} + BB^*K \text{ is stable}\}$. Then Φ is an open subset of the vector space of all self-adjoint n by n matrices K. We define a map $f: \Phi \to \{\text{all } n \text{ by } n \text{ self-adjoint matrices}\}$ as follows. Given K in Φ, set $f(K) = L$, where L is the unique solution of

$$g(K; L) = A^*L + LA + C^*C = 0, \qquad A = \bar{A} + BB^*K.$$

By using the implicit function theorem (as in the proof of 2.5.4) one shows that the map f is C^∞.

2.5.14. Exercise. Let $\alpha \geq 0$. Suppose $K > 0$ satisfies (5.6). Show that $K \leq L^{-1}$, where $L > 0$ solves $\bar{A}L + L\bar{A}^* + BB^* = 0$. (*Hint:* See 1.2.14.)

2.5.15. Exercise. Suppose that $K(\alpha)$, $0 \leq \alpha < \alpha_1$, lies in Φ and solves the nonlinear differential equation

$$\frac{dK}{d\alpha} = f(K) \qquad \text{with} \quad K(0) = 0. \tag{5.9}$$

Then $K(\alpha) \geq 0$ and solves (5.6) for $0 \leq \alpha < \alpha_1$.

2.5.16. Proposition. *The initial value problem* (5.9) *has a unique solution* $K(\alpha)$, $0 \leq \alpha < 1$, *in* Φ.

PROOF. Since f is C^∞, the solution is unique as long as it exists. Since 0 is in Φ, the solution exists for small positive α and lies in Φ.

Let α_1 be the first α for which $K(\alpha)$ leaves Φ: $\alpha_1 = \sup\{\alpha | K(\alpha) \text{ in } \Phi\}$. Claim: $\alpha_1 \geq 1$. Suppose not; suppose that $\alpha_1 < 1$. Then, since $f(K) \geq 0$, $K(\alpha)$ is an increasing function of α. By 2.5.14, $K(\alpha) \leq L^{-1}$ and so the limit $K(\alpha_1)$ of $K(\alpha)$ as $\alpha \to \alpha_1$ exists. By 2.5.15, $K(\alpha)$ satisfies (5.6) and hence $K(\alpha_1)$ also satisfies (5.6) with $\alpha = \alpha_1$. Thus $K(\alpha)$, $0 \leq \alpha \leq \alpha_1$, is nonnegative and satisfies (5.6). By 2.5.13, the eigenvalues of $A(\alpha)$ are therefore not on the imaginary axis. Since the eigenvalues are continuous functions of α and they are stable for $\alpha = 0$, we have that $A(\alpha)$ is stable for all $\alpha \leq \alpha_1$. In particular, $K(\alpha_1)$ is in Φ, contradicting the definition of α_1. Thus $\alpha_1 \geq 1$. □

2.5.17. Corollary. *There exists a* $K > 0$

$$\bar{A}^*K + K\bar{A} + KBB^*K + C^*C = 0, \tag{5.10}$$

with the eigenvalues λ *of* $A = \bar{A} + BB^*K$ *satisfying* $\text{Re}(\lambda) \leq 0$.

PROOF. Let $K(\alpha)$, $0 \leq \alpha < 1$, be the solution of (5.9). Then, by 2.5.15, $K(\alpha)$ satisfies (5.6). Thus, by letting $\alpha \to 1$, we see that $K = K(1) \leq L^{-1}$ exists and satisfies (5.10). The rest follows from 2.5.10 and the fact that $A(\alpha)$ is stable for $\alpha < 1$. □

We can now prove Theorem 2.5.8. Let $\bar{G}(s)$ be a bounded real transfer function and let $(\bar{A}, B, C)$ be a minimal realization. Let A be as in 2.5.17. Then (5.10) becomes

$$\bar{A}^*K + KA + C^*C = 0.$$

Since $\bar{A}$ is stable and $\text{Re}(\lambda) \leq 0$ for all eigenvalues λ of A, we can write

$$K = \int_0^\infty e^{t\bar{A}^*} C^* C e^{tA} \, dt, \tag{5.11}$$

with the integral absolutely convergent. Suppose now that $CA^k x = 0$, $k = 0, 1, \ldots$. Then we have $Kx = 0$ and so $x^*Kx = 0$. Since $K > 0$, this forces $x = 0$ and shows that (A, C) is observable. Clearly (A, B) is controllable; thus (A, B, C) is minimal. Now (5.10) can be rewritten as

$$A^*K + KA = KBB^*K - C^*C.$$

Thus $K = K(A, B, C)$ and so $\bar{A} = A - BB^*K$. This concludes the proof of Theorem 2.5.8.

2.5.18. Corollary. *Let $\bar{G}(s)$ be a bounded real p by m transfer function. Then there is an m by m stable transfer function $H(s)$ such that*

$$H(-s^*)^*H(s) = I - \bar{G}(-s^*)^*\bar{G}(s). \tag{5.12}$$

PROOF. Let $\bar{G}^\#(s)$ be given by (5.7) with $\alpha = 1$ and set $H(s) = I - \bar{G}^\#(s)$. By 2.5.8 and 2.3.10, $H(s)$ exists and (5.12) holds. □

2.5.19. Exercise. Suppose K_i, $i = 1, 2$, are both self-adjoint solutions of (5.10). Suppose that $A_i = \bar{A} + BB^*K_i$, $i = 1, 2$, are both stable. Then $K_1 = K_2$.

A transfer function $\bar{G}(s)$ is *strictly bounded real* if strict inequality holds in the definition: $I - \bar{G}(i\omega)^*\bar{G}(i\omega) > 0$ for all real ω. A triple $(\bar{A}, B, C)$ is said to be (strictly) bounded real if the corresponding transfer function is so.

2.5.20. Corollary. *The map $(A, B, C) \to (\bar{A}, B, C)$ is a bijection between the space of all stable minimal systems and the space of all strictly bounded real minimal systems. Moreover, this map is C^∞ and its inverse is C^∞.*

PROOF. By 2.5.7, we know that $Q(\omega) = I - \bar{G}(i\omega)^*\bar{G}(i\omega) \geq 0$ for all real ω. Suppose now that $v^*Q(\omega)v = 0$ for some ω and some nonzero v. Then, by 2.3.10 and 2.3.11, $i\omega$ is a pole of $I + G^\#(s)$. This implies that $i\omega$ is an eigenvalue of A, which by assumption cannot hold. Thus $\bar{G}(s)$ is strictly bounded real. We show now that $(\bar{A}, C)$ is observable. Indeed, if $\bar{A}x = \lambda x$ and $Cx = 0$, then, by (5.11), $x^*K = 0$ and so $x = 0$. Thus all eigenvalues of $\bar{A}$ are observable; since $(\bar{A}, B)$ is clearly controllable, 1.5.7 implies that $(\bar{A}, C)$ is observable. Conversely, let $(\bar{A}, B, C)$ be a strictly bounded real system and let A be as in 2.5.17.

If A had an eigenvalue on the imaginary axis, then, by 2.5.11, with $\alpha = 1$, $G^\#(s)$ would have a pole on the imaginary axis; this would, by 2.3.11, contradict the strict bounded realness of $(\bar{A}, B, C)$. Now suppose that (A_1, B, C) and (A_2, B, C) correspond to the same $(\bar{A}, B, C)$. Let $K_i = K(A_i, B, C)$; then K_i, $i = 1$, 2, both satisfy (5.10) and so, by 2.5.19, $K_1 = K_2$. Thus $A_1 = \bar{A} + BB^*K_1 = \bar{A} + BB^*K_2 = A_2$. This shows that the map is a bijection. Smoothness follows from 2.5.4 while smoothness of the inverse map follows by applying a similar argument to (5.10).   □

In closing, it should be pointed out that property (ii) in the definition of bounded real is, up to a scale factor, implied by property (i), as the following exercise shows.

2.5.21. Exercise. If $\bar{G}(s)$ is stable then there is an $\varepsilon > 0$ such that $\varepsilon\bar{G}(s)$ is strictly bounded real.

In particular, 2.5.20 and 2.5.21 combined imply that for any stable transfer function $\bar{G}(s)$, $\varepsilon\bar{G}(s)$ is an optimally controlled transfer function for some $\varepsilon > 0$.

2.6. Notes and References

The *LQ* regulator problem was initially posed and solved by Kalman in 1960 [2.3]. The solution presented is classical and the development follows that appearing in [1.1] with one exception, the emphasis on the concept of properness. This is included here because of the fact that its usefulness carries over to certain nonlinear extensions. Specifically, if $f: \mathbb{R}^n \to \mathbb{R}^n$, $g: \mathbb{R}^n \to \mathbb{R}^{n \times m}$, and $h: \mathbb{R}^n \to \mathbb{R}^p$ are smooth, consider the nonlinear control system

$$\dot{x} = f(x) + g(x)u, \qquad x(0) = x^0 \text{ in } \mathbb{R}^n,$$

$$y = h(x),$$

where we restrict ourselves to the real situation. Let $J^u(x^0)$ and $S(x^0)$ be as in the chapter and assume that $f(0) = 0$ and $h(0) = 0$. The triple (f, g, h) is said to be *proper* if S is so; (f, h) is *observable* if $u = 0$ and $y^u(\cdot; x^0) = y^u(\cdot; 0)$ iff $x^0 = 0$. Then 2.2.8 holds with essentially the same proof. The triple (f, g, h) is *homogeneous of degree r* if f and h are so while g is homogeneous of degree zero. By modifying the proof of 2.2.9, it can be checked that this implies that S is homogeneous of degree $r + 1$. When $r > -1$, this homogeneity can be used as in 2.2.10 to conclude that S is proper. The results of Section 2.2 thus carry over naturally to a nonlinear setting. It must be cautioned that at present very little is known about the nonlinear extension of the algebraic Riccati equation.

Much of Section 2.5 is due to D. F. Delchamps. Specifically, the smoothness of K (2.5.4) and the main result (2.5.20) are due to him [2.1]. The crucial 2.5.8

appears in [1.1]; the proof presented here is due to Delchamps (private communication). The implicit function theorem can be found in Rudin [2.7]. The existence theorem for nonlinear ordinary differential equations can be found in Hirsch and Smale [1.4].

A thorough investigation of the Riccati differential equation is [2.8]. The proof of Lemma 2.2.1 is in [2.6], p. 128. A nice introduction to the study of the geometry of $M(m, n, p)$ is [2.9]. The references therein provide additional material. From the viewpoint of the general theory of nonlinear control, the *LQ* regulator is a very special case. Presentations of nonlinear theory are [2.2], [2.4], and [2.5].

[2.1]   D. F. Delchamps, "Analytic Feedback Control and the Algebraic Riccati Equation," *IEEE Trans. Automat. Control*, **29** (1984), 1031–1033.
[2.2]   W. H. Fleming and R. Rishel, *Deterministic and Stochastic Optimal Control*, Springer-Verlag, New York, 1975.
[2.3]   R. E. Kalman, "Contributions to the Theory of Optimal Control," *Bol. Soc. Mat. Mexicana* (1960).
[2.4]   N. V. Krylov, *Controlled Diffusion Processes*, Springer-Verlag, New York, 1980.
[2.5]   P. L. Lions, *Generalized Solutions of Hamilton–Jacobi Equations*, Pitman, London, 1982.
[2.6]   D. Luenberger, *Optimization by Vector Space Methods*, Wiley, New York, 1969.
[2.7]   W. Rudin, *Principles of Mathematical Analysis*, 2nd edn., McGraw-Hill, New York, 1964.
[2.8]   M. Shayman, "Phase Portrait of the Matrix Riccati Equation," *SIAM J. Control Optim.*, **24** (1986), 1–65.
[2.9]   A. Tannenbaum, *Invariance and System Theory: Algebraic and Geometric Aspects*, Lecture Notes in Mathematics 845, Springer-Verlag, New York, 1981.

CHAPTER 3

Brownian Motion

3.1. Preliminary Definitions

Until now our control systems have been defined over the complex numbers so as to simplify the linear algebra. Since we have no more need to do so, and since Brownian motion is more familiar over the real numbers, from now on we work exclusively in the real domain: henceforth all vectors and matrices will have real values.

3.1.1. Exercise. Let (A, B, C) be a minimal triple that is real. Show that $K = K(A, B, C)$ is real (Section 2.5).

The purpose of Chapter 5 is to describe a probabilistic extension of the LQ regulator and related topics. The present chapter is a lengthy digression on the required probabilistic foundations.

An *event space* is a pair $(\Omega, \mathscr{F})$, where $\mathscr{F}$ is a σ-algebra on the set Ω. Every *metric* space X can be made into an event space by setting $\mathscr{F} = \mathscr{B}(X)$, the *Borel* σ-algebra on X; this is the smallest σ-algebra containing all the open sets in X. In particular, Euclidean space $\mathbb{R}^m$ will always come equipped with its standard Borel σ-algebra $\mathscr{B}(\mathbb{R}^m)$ induced by the usual Euclidean metric. Any set Ω can be trivially made into a metric space by setting $d(\omega, \omega') = 1$ if $\omega \neq \omega'$ and $d(\omega, \omega) = 0$ for all ω, ω' in Ω. Then all subsets of Ω are open and so the collection 2^Ω of all subsets of Ω is the corresponding σ-algebra. The *product* of event spaces $(\Omega, \mathscr{F})$, $(\Omega', \mathscr{F}')$ is the event space $(\Omega \times \Omega', \mathscr{F} \times \mathscr{F}')$, where $\mathscr{F} \times \mathscr{F}'$ is the smallest σ-algebra containing all sets of the form $A \times A'$, for A in $\mathscr{F}$ and A' in $\mathscr{F}'$. A map $x: \Omega \to X$ is *$\mathscr{F}$-measurable* if $x^{-1}(G)$ is in $\mathscr{F}$ whenever G is open in X. A *process on* Ω is a map on the product

$x: [0, \infty) \times \Omega \to X$; equivalently, a process is a family of maps $x(t): \Omega \to X$, $t \geq 0$.

3.1.2. Exercise. Show that $x: [0, \infty) \times \Omega \to X$ is $\mathscr{B}([0, \infty)) \times 2^{\Omega}$ measurable implies that for all ω in Ω, $x(\cdot, \omega): [0, \infty) \to X$ is $\mathscr{B}([0, \infty))$ measurable.

A family of σ-algebras $\mathscr{F}_t$, $t \geq 0$, on a set Ω is *nondecreasing* if $\mathscr{F}_s \subset \mathscr{F}_t$ whenever $s < t$. A process $x(\cdot)$ is $\mathscr{F}_t$-*progressively* *measurable* if the map x restricted to $[0, T] \times \Omega$ is $\mathscr{B}([0, T]) \times \mathscr{F}_T$ measurable for all $T \geq 0$. By 3.1.2, it then follows that the process has measurable sample paths $x(\cdot, \omega)$, ω in Ω. Given a process $x(\cdot)$, let $\mathscr{X}_t = \sigma(x(s), 0 \leq s \leq t)$ denote the smallest σ-algebra on Ω relative to which the maps $x(s)$, $0 \leq s \leq t$, are measurable. $\mathscr{X}_t$, $t \geq 0$, is the family of σ-algebras *generated* by the process $x(\cdot)$. In general, the process $x(\cdot)$ need not be $\mathscr{X}_t$-progressively measurable.

For any set $B \subset \Omega$, let $1_B(\omega) = 1$ if $\omega \in B$ and $= 0$ if $\omega \notin B$. Let $a \wedge b = \min(a, b)$ for any $-\infty \leq a, b \leq +\infty$. A function of time $\alpha: [0, \infty) \to X$ is *right continuous* if the limit $\alpha(t+) = \lim_{s \downarrow 0} \alpha(t + s)$ exists and equals $\alpha(t)$ for all $t \geq 0$. A process is *right continuous* if *all* its sample paths are so. Let $\mathscr{F}_t$, $t \geq 0$, be a nondecreasing family of σ-algebras on Ω. A process $x(\cdot)$ is $\mathscr{F}_t$-*nonanticipating* if $x(t)$ is $\mathscr{F}_t$-measurable for all $t \geq 0$. Clearly, progressive measurability implies nonanticipativeness. A map $\tau: \Omega \to [0, \infty) \cup \{\infty\}$ is an $\mathscr{F}_t$-*stopping time* if $\{\tau \leq t\} \subset \Omega$ is in $\mathscr{F}_t$ for all $t \geq 0$.

3.1.3. Exercises. Let $x(\cdot)$ be nonanticipating and right continuous. Then $x(\cdot)$ is progressively measurable. (*Hint:* Fix $T \geq 0$ and set $x_n(t, \omega) = x((kT/n) \wedge T, \omega)$ whenever $(k - 1)T \leq nt < kT, k \geq 1$. Show that $x_n(\cdot)$ is measurable over $\mathscr{B}([0, T]) \times \mathscr{F}_T$ and hence $x(\cdot)$ is.)

Applying 3.1.3, we see that with $\mathscr{X}_t$ as above and $x(\cdot)$ right continuous, the process $x(\cdot)$ is progressively measurable relative to $\mathscr{X}_t$. More generally set $x_T(t, \omega) = x(t \wedge T, \omega)$; then for all $T \geq 0$ the process $x_T(\cdot)$ is $\mathscr{X}_t$-progressively measurable. Let us say that a subset $A \subset [0, \infty) \times \Omega$ is progressively measurable if the process 1_A is so. Then the collection of progressively measurable sets forms a σ-algebra on $[0, \infty) \times \Omega$ and a process is progressively measurable if and only if it is measurable relative to the σ-algebra of progressively measurable sets. In particular, let F be a jointly measurable map on the infinite product $[0, \infty) \times X \times X \times \cdots$ and let t_k, $k \geq 1$, be a fixed sequence of nonnegative numbers. Then

$$z(t, \omega) = F(t, x(t \wedge t_1, \omega), x(t \wedge t_2, \omega), \ldots), \qquad t \geq 0, \quad \omega \text{ in } \Omega,$$

is an $\mathscr{X}_t$-progressively measurable process. It turns out that when $x(\cdot)$ is right continuous, *any* progressively measurable process $z(\cdot)$ can be expressed as above for some sequence $t_k \geq 0$, $k \geq 1$, and measurable F.

Note also that, by 3.1.3, the process $1_{t < \tau}$, $t \geq 0$, is progressively measurable whenever τ is a stopping time. The only stopping time used in this book is the following.

3.1.4. Exercise. Let $x(\cdot)$ be progressively measurable and right continuous and let C be closed. Show that the *contact time*

$$\tau = \inf\{t \geq 0 | \overline{(x(s), 0 \leq s \leq t)} \cap C \neq \phi\}$$

is a stopping time.

3.1.5. Exercise. Let $x(\cdot)$ be as in 3.1.4 and suppose ω is such that $x(\cdot, \omega)$ is continuous. Show that the contact time $\tau(\omega)$ and the *entrance time* $\tau_C(\omega)$ agree, where

$$\tau_C = \inf\{t \geq 0 | x(t) \in C\}.$$

Let $z(\cdot)$ be a bounded $\mathbb{R}^m$-valued progressively measurable process, $|z(t, \omega)| \leq$ constant for all t and ω. Then $\int_0^t |z(s)|^2 \, ds, t \geq 0$, is nonanticipating and right continuous and thus is progressively measurable. Since the sample paths are continuous, 3.1.5 implies that

$$\tau_n = \inf\left\{T > 0 \middle| \int_0^T |z(t)|^2 \, dt \geq n\right\}$$

is a stopping time. Let $g(\cdot)$ be progressively measurable and let $c > 0$. Then

$$g_c(t) = 1\left(\int_0^t |g(s)|^2 \, ds \leq c\right) g(t), \qquad t \geq 0,$$

is again progressively measurable. This can be seen by noting that

$$\left\{\int_0^t |g(s)|^2 \, ds \leq c\right\} = \bigcap_{n \geq 1}\left\{\int_0^t n \wedge |g(s)|^2 \, ds \leq c\right\};$$

since $n \wedge |g(\cdot)|^2$ is bounded, the result follows from the above. Note that

$$\int_0^\infty |g_c(t)|^2 \, dt \leq c$$

for *all* ω.

A *probability space* is a triple $(\Omega, \mathscr{F}, P)$, where $(\Omega, \mathscr{F})$ is an event space and P is a *probability measure* on $(\Omega, \mathscr{F})$. A *random variable* on $(\Omega, \mathscr{F}, P)$ is an $\mathscr{F}$-measurable map $x: \Omega \to X$. A real random variable $x: \Omega \to \mathbb{R}$ is *integrable* if the *expectation* $E^P(|x|)$ is finite. For each $r > 0$, the set of all random variables whose rth power is integrable is denoted $L^r = L^r(\Omega, \mathscr{F}, P)$. A process $x(\cdot)$ is *P-almost surely continuous* if the set Ω' of all ω's for which $x(\cdot, \omega)$ is continuous has measure one, $P(\Omega') = 1$.

For convenience we let $L^0 = L^0(\Omega, \mathscr{F}, P)$ be the set of all random variables. We say that a sequence $x_n: \Omega \to \mathbb{R}, n \geq 1$, converges to x *in probability* if for all $\varepsilon > 0$, $P(|x_n - x| \geq \varepsilon) \to 0$ as $n \uparrow \infty$. This is also called *convergence in L^0*.

The sequence *converges in* L^r, $r > 0$, if $E^P(|x_n - x|^r) \to 0$ as $n \uparrow \infty$. In particular, convergence in L^1 is also called *convergence in the mean.* Since

$$P(|x_n - x| \geq \varepsilon) \leq \frac{1}{\varepsilon^r} E^P(|x_n - x|^r) \tag{1.1}$$

for all $\varepsilon > 0$, convergence in L^r, $r > 0$, implies convergence in probability.

It is well known that the spaces L^r, $r \geq 0$, are *complete.* A basic and frequently used fact is the corresponding result in the calculus of stochastic processes:

Completeness Lemma. *Let* x_n: $[0, \infty) \times \Omega \to \mathbb{R}^d$ *be progressively measurable right continuous processes, $n \geq 1$. Suppose that*

$$\lim_{n \to \infty} \sup_{m \geq n} P\left(\sup_{0 \leq t \leq T} |x_n(t) - x_m(t)| \geq \varepsilon \right) = 0$$

for all $T > 0$ *and all* $\varepsilon > 0$. *Then there is a progressively measurable right continuous process* x: $[0, \infty) \times \Omega \to \mathbb{R}^d$ *satisfying*

$$\lim_{n \to \infty} P\left(\sup_{0 \leq t \leq T} |x_n(t) - x(t)| \geq \varepsilon \right) = 0$$

for all $T > 0$ *and all* $\varepsilon > 0$.

Analogous statements for convergence in L^r, $r > 0$, hold. Since we need only this lemma, they are omitted. We emphasize that this lemma is nontrivial, due to the fact that $\mathscr{F}_t$, $t \geq 0$, is arbitrary (noncomplete).

3.1.6. Exercise. Show that if x_n are *uniformly bounded* random variables ($|x_n(\omega)| \leq$ constant for all n and ω) then convergence in probability implies convergence in L^r for all $r > 0$.

More generally, a sequence $\{x_n\}$ is *uniformly integrable* if

$$\sup_{n \geq 1} E^P(|x_n|; |x_n| \geq a) \to 0 \qquad \text{as} \quad a \uparrow \infty.$$

In this case convergence in probability implies convergence in the mean.

Let $(\Omega, \mathscr{F}, P)$ be a probability space and let R be an integrable real random variable. Define Q on $(\Omega, \mathscr{F})$ by setting

$$Q(A) = E^P(R; A), \qquad A \text{ in } \mathscr{F}, \tag{1.2}$$

the expectation of R over the set A. Then Q is a measure on $(\Omega, \mathscr{F})$ and Q is *absolutely continuous* with respect to P, i.e., $Q(N) = 0$ whenever $P(N) = 0$. If $P(R > 0) = 1$, then P is also absolutely continuous relative to Q; in this case one says that P and Q are mutually absolutely continuous. More generally, if Q is any finite measure on $(\Omega, \mathscr{F})$ which is absolutely continuous with respect to P, there exists an integrable real random variable R such that (1.2) holds. Such an R is determined P-almost surely and is denoted $R = dQ/dP$. Absolute continuity is denoted as $Q \ll P$.

3.1.7. Exercise. Let $Q \ll P$. Suppose A_n satisfies $P(A_n) \to 0$. Show that $Q(A_n) \to 0$. Conclude that if x_n converges in P-probability then x_n converges in Q-probability.

3.1.8. Exercise. Continuing 3.1.7, show that if a uniformly bounded sequence converges in $L'(\Omega, \mathscr{F}, P)$ then it converges in $L'(\Omega, \mathscr{F}, Q)$.

Let $(\Omega, \mathscr{F}, P)$ be a probability space and let $\mathscr{D}$ be a sub-σ-algebra of $\mathscr{F}$. Let R be integrable and let Q be given by (1.2). Let P', Q' denote the restrictions of P, Q to $\mathscr{D}$. Then $Q' \ll P'$ and hence the $\mathscr{D}$-measurable integrable random variable dQ'/dP' exists. This random variable is denoted $E^P(R|\mathscr{D})$ and is the *conditional expectation* of R *given* $\mathscr{D}$. Thus $f = E^P(R|\mathscr{D})$ a.s. P if f is $\mathscr{D}$-measurable and $E^P(f; A) = E^P(R; A)$ for all A in $\mathscr{D}$. In particular, $P(B|\mathscr{D}) = E^P(1_B|\mathscr{D})$ is the *conditional probability* of B *given* $\mathscr{D}$.

Let $\mathscr{D}_1$, $\mathscr{D}_2$ be sub-σ-algebras of $\mathscr{F}$. $\mathscr{D}_1$ and $\mathscr{D}_2$ are *independent* (under P) if $P(A_1 \cap A_2) = P(A_1)P(A_2)$ for all A_1 in $\mathscr{D}_1$ and A_2 in $\mathscr{D}_2$. Random variables x and y are *independent* if $\sigma[x]$ and $\sigma[y]$ are independent. Processes $x(\cdot)$ and $y(\cdot)$ are *independent* if $\sigma[x(t), t \geq 0]$ and $\sigma[y(t), t \geq 0]$ are independent.

Let $\mathscr{F}_t, t \geq 0$, be a nondecreasing family of sub-σ-algebras on a probability space $(\Omega, \mathscr{F}, P)$. An $\mathbb{R}^m$-valued progressively measurable right continuous process $\eta: [0, \infty) \times \Omega \to \mathbb{R}^m$ is an $(\Omega, \mathscr{F}_t, P)$ *Brownian motion* if

(i) $\eta(\cdot)$ is P-almost surely continuous,
(ii) $P(\eta(0) = 0) = 1$,
(iii) $P(\eta(t) \in B|\mathscr{F}_s) = \int_B g_m(t - s, x - \eta(s))\, dx$ a.s. P for all $0 \leq s < t$ and B in $\mathscr{B}(\mathbb{R}^m)$.

Here $g_m: (0, \infty) \times \mathbb{R}^m \to \mathbb{R}$ is given by $g_m(t, x) = (2\pi t)^{-m/2} \exp(-|x|^2/2t)$. The function g_m is the Gauss–Weierstrass kernel.

3.1.9. Exercise. Let $\eta(\cdot)$ be an $(\Omega, \mathscr{F}_t, P)$ Brownian motion. Show that $\sigma[\eta(t), t \geq 0]$ is independent of $\mathscr{F}_0$.

A progressively measurable right continuous process $R(\cdot)$ is a $(\Omega, \mathscr{F}_t, P)$ *martingale* if:

(i) $R(\cdot)$ is P-almost surely continuous,
(ii) $R(t)$ is integrable for all $t \geq 0$,
(iii) $E^P(R(t)|\mathscr{F}_s) = R(s)$ a.s. P for all $t > s \geq 0$.

Usually when property (i) is assumed, $R(\cdot)$ is called a "continuous martingale." We do not do so here as our martingales will always satisfy (i).

3.1.10. Exercise. Let $\eta(\cdot)$ be progressively measurable and right continuous. Show that $\eta(\cdot)$ is a Brownian motion iff $P(\eta(0) = 0) = 1$ and $R(t) = \exp(c\eta(t) - |c|^2 t/2), t \geq 0$, is a martingale for all row vectors c in $\mathbb{R}^m$.

3.1.11. Exercise. Let f be an integrable random variable. Show that f is independent of a sub-σ-algebra $\mathscr{D}$ iff $E^P(f|\mathscr{D}) = E^P(f)$ a.s. P.

Let $N \geq 1$ be an integer and let $\theta: \Omega \to \{1, \ldots, N\}$ be a random variable on $(\Omega, \mathscr{F}, P)$. Given an N-tuple of real numbers $\pi = (\pi_1, \ldots, \pi_N)$ recall that θ is *distributed according to* π if $P(\theta = j) = \pi_j, j = 1, \ldots, N$. In this case $0 \leq \pi_j \leq 1$ and $\pi_1 + \cdots + \pi_N = 1$.

In what follows we will be working with a probability space $(\Omega, \mathscr{F}, P)$ *equipped with* a nondecreasing family of sub-σ-algebras $\mathscr{F}_t$, $t \geq 0$, of $\mathscr{F}$, an $(\Omega, \mathscr{F}_t, P)$ Brownian motion $\eta(\cdot)$ in $\mathbb{R}^m$, and, for each $N \geq 1$ and distribution π on $\{1, \ldots, N\}$, a random variable $\theta = \theta(\cdot, N, \pi): \Omega \to \{1, \ldots, N\}$ that is $\mathscr{F}_0$-measurable and is distributed according to π. By 3.1.9, this implies that θ and $\eta(\cdot)$ are independent. To establish that there is no problem concerning the *existence* of such a space, we outline a construction that yields such a space. As this construction is not necessary for what follows, the reader may wish to skip the following discussion.

Let $C([0, \infty); \mathbb{R}^m)$ denote the set of all continuous paths $\alpha: [0, \infty) \to \mathbb{R}^m$. Let $b(t): C([0, \infty); \mathbb{R}^m) \to \mathbb{R}^m$ be given by $b(t, \alpha) = \alpha(t)$. For each $T \geq 0$, let $\mathscr{M}_T = \sigma[b(t), 0 \leq t \leq T]$. Let $\mathscr{M}$ denote the smallest σ-algebra containing all the $\mathscr{M}_T$'s. $C([0, \infty); \mathbb{R}^m)$ is a metric space in a natural manner; $\mathscr{M}$ is then the Borel σ-algebra of this metric space. There *exists* a probability measure W on the event space $(C([0, \infty); \mathbb{R}^m), \mathscr{M})$ uniquely characterized by the statement that the process $b(\cdot)$ is a $(C([0, \infty); \mathbb{R}^m), \mathscr{M}_t, W)$ Brownian motion. This measure, originally discovered by N. Wiener in 1923, is referred to as the *Wiener measure*. In this context, the path space $C([0, \infty); \mathbb{R}^m)$ is referred to as the *Wiener space. The existence of the Wiener measure lies at the basis of much of probability theory and in particular lies at the basis of all that follows.*

Let $\Omega = C([0, \infty); \mathbb{R}^m) \times [0, 1]$. Denote elements of Ω by $\omega = (\alpha, \beta)$ with $0 \leq \beta \leq 1$. Let $\mathscr{F}_t = \mathscr{M}_t \times \mathscr{B}([0, 1])$, $\mathscr{F} = \mathscr{M} \times \mathscr{B}([0, 1])$, and $P = W \times$ {Lebesgue measure} on $(\Omega, \mathscr{F})$. P is the unique measure on $\mathscr{F}$ satisfying $P(A \times [\beta_1, \beta_2]) = W(A) \cdot (\beta_2 - \beta_1)$ for all A in $\mathscr{M}$ and $0 \leq \beta_1 \leq \beta_2 \leq 1$. Define $\eta(t, \omega) = \alpha(t)$, $t \geq 0$, and for each $N \geq 1$ and distribution π on $\{1, \ldots, N\}$, define $\theta(\omega) = 1$ if $0 \leq \beta \leq \pi_1$, $\theta(\omega) = 2$ if $\pi_1 \leq \beta < \pi_1 + \pi_2, \ldots,$ $\theta(\omega) = N$ if $\pi_1 + \cdots + \pi_{N-1} \leq \beta \leq 1$. This completes the construction.

3.1.12. Exercise. Let $(\Omega, \mathscr{F}, P)$ be a probability space and let $\mathscr{D}$ be a sub-σ-algebra of $\mathscr{F}$. Let f be an integrable random variable, let g be a $\mathscr{D}$-measurable random variable, and suppose that fg is integrable. Show that $E^P(fg|\mathscr{D}) = E^P(f|\mathscr{D})g$ a.s. P.

3.1.13. Exercise. Let $\eta(\cdot)$ be an $(\Omega, \mathscr{F}_t, P)$ Brownian motion. Show that $c\eta(t)$, $t \geq 0$, and $\eta(t)^*Q\eta(t) - \text{trace}(Q)t$, $t \geq 0$, are $(\Omega, \mathscr{F}_t, P)$ martingales. Here c is a row vector and $Q = Q^*$. (*Hint:* Replace c by εc in 3.1.10 and differentiate with respect to ε.)

3.1.14. Exercise. Suppose that for each $T \geq 0$, $x_T \colon [0, \infty) \times \Omega \to X$ is progressively measurable and right continuous with

$$P(x_T(t) = x_{T'}(t), 0 \leq t \leq T) = 1,$$

whenever $T < T'$. Show that there is a unique progressively measurable right continuous process $x(\cdot)$ satisfying

$$P(x(t) = x_T(t), 0 \leq t \leq T) = 1$$

for all $T \geq 0$.

A variant of the previous exercise is the following.

3.1.15. Exercise. Let $0 \leq \tau_1 \leq \tau_2 \leq \cdots$ be a nondecreasing sequence of stopping times such that $P(\tau_n \uparrow \infty \text{ as } n \uparrow \infty) = 1$. Suppose that for each $n \geq 1$, $x_n(\cdot)$ is progressively measurable right continuous. Assume that the processes $x_n(\cdot)$ are consistently defined: $P(x_{n+1}(t) = x_n(t), 0 \leq t < \tau_n) = 1$ for all $n \geq 1$. Show that there exists a P-almost surely unique progressively measurable right continuous process $x(\cdot)$ satisfying $P(x(t) = x_n(t), 0 \leq t < \tau_n) = 1$.

3.1.16. Exercise. Let $\eta(\cdot)$ be an $(\Omega, \mathscr{F}_t, P)$ Brownian motion and let $\mathscr{D}_t, t \geq 0$, be a nondecreasing family of σ-algebras satisfying $\sigma[\eta(s), 0 \leq s \leq t] \subset \mathscr{D}_t \subset \mathscr{F}_t$ for all $t \geq 0$. Show that $\eta(\cdot)$ is an $(\Omega, \mathscr{F}_t, P)$ Brownian motion.

The following deals with an important inequality of J. L. Doob's that holds for martingales. Apart from its intrinsic interest, the inequality is crucial for the construction of the stochastic integral (next section).

3.1.17. Lemma. *Let $\mathscr{F}_1 \subset \mathscr{F}_2 \subset \cdots \subset \mathscr{F}_N$ be sub-σ-algebras of $\mathscr{F}$ and let $x_1, x_2, \ldots, x_N$ be real integrable random variables on $(\Omega, \mathscr{F}, P)$ with $x_k \mathscr{F}_k$-measurable, $k = 1, \ldots, N$, and satisfying $E^P(x_n | \mathscr{F}_m) \geq x_m$ whenever $m < n$. Then, for all $1 \leq n \leq N$,*

$$P\left(\max_{1 \leq k \leq n} x_k \geq \lambda \right) \leq \frac{1}{\lambda} E^P\left(x_n; \max_{1 \leq k \leq n} x_k \geq \lambda \right) \tag{1.3}$$

for all positive λ.

PROOF. Let B_n be the event whose probability is on the left-hand side of (1.3). Set $C_1 = B_1$, $C_2 = B_2 \cap B_1^c$, $C_3 = B_3 \cap B_2^c$, $\ldots$. Then C_j are disjoint, C_j is $\mathscr{F}_j$-measurable, and their union $C_1 \cup C_2 \cup \cdots \cup C_n$ equals B_n. Note that on C_j, $x_j \geq \lambda$. Let $\sum$ denote summation from 1 to n. Then

$$P(B_n) = \sum P(C_j) \leq \sum \frac{1}{\lambda} E^P(x_j; C_j) \leq \sum \frac{1}{\lambda} E^P(x_n; C_j) = \frac{1}{\lambda} E^P(x_n; B_n). \qquad \square$$

3.1.18. Exercise. Let $x(\cdot)$ be an $(\Omega, \mathscr{F}_t, P)$ martingale. Show that

$$P\left(\sup_{0 \leq t \leq T} |x(t)| \geq \lambda \right) \leq \frac{1}{\lambda} E^P\left(|x(T)|; \sup_{0 \leq t \leq T} |x(t)| \geq \lambda \right) \tag{1.4}$$

for all $\lambda > 0$ and $T \geq 0$. (*Hint*: Use the fact that $x(\cdot)$ is right continuous and apply 3.1.17.)

3.1.19. Exercise. Let f, g be nonnegative random variables on $(\Omega, \mathcal{F}, P)$. Assume that $P(f \geq \lambda) \leq (1/\lambda) E^P(g; f \geq \lambda)$, $\lambda > 0$. Show that

(a)
$$E^P(f^r) = r \int_0^\infty \lambda^{r-1} P(f \geq \lambda) \, d\lambda \qquad (r > 0),$$

(b)
$$E^P(f^r) \leq \left(\frac{r}{r-1}\right)^r E^P(g^r) \qquad (r > 1).$$

3.1.20. Exercise. Let $x(\cdot)$ be an $(\Omega, \mathcal{F}_t, P)$ martingale. Show that

$$E^P\left(\sup_{0 \leq t \leq T} |x(t)|^r\right) \leq \left(\frac{r}{r-1}\right)^r E^P(|x(T)|^r) \qquad (r > 1);$$

in particular,

$$E^P\left(\sup_{0 \leq t \leq T} |x(t)|^2\right) \leq 4 E^P(|x(T)|^2). \tag{1.5}$$

We use (1.5) to derive Doob's *martingale convergence theorem*. This will be useful in Section 4.1.

Let $x(\cdot)$ be an $\mathbb{R}^n$-valued $(\Omega, \mathcal{F}_t, P)$ martingale such that $E^P(|x(T)|^2)$ is a bounded function of $T \geq 0$. Note that the a.s. continuity of $x(\cdot)$ is unused in the derivation of (1.8) and (1.9).

3.1.21. Exercise. Show that $E^P(|x(\cdot)|^2)$ is a nondecreasing function of time and that

$$\lim_{s \uparrow \infty} \sup_{t \geq s} E^P(|x(t) - x(s)|^2) = 0. \tag{1.6}$$

(*Hint*: Expand the left-hand side of (1.6).)

3.1.22. Exercise. Show that

$$P\left(\sup_{s \leq t \leq T} |x(t) - x(s)| \geq \varepsilon\right) \leq \frac{1}{\varepsilon} E^P(|x(T) - x(s)|^2)^{1/2}.$$

Conclude that

$$P\left(\limsup_{s \to \infty} \sup_{t \geq s} |x(t) - x(s)| \geq \varepsilon\right) = 0 \tag{1.7}$$

for all $\varepsilon > 0$.

Thus, by (1.7), the sequence $x(t, \omega)$, $t \geq 0$ is Cauchy for P-almost all ω. Hence

$$P\left(\lim_{t \to \infty} x(t) = x(\infty) \text{ exists}\right) = 1. \tag{1.8}$$

Now since $E^P(|x(\infty) - x(s)|^2) \leq \lim \inf_{t \to \infty} E^P(|x(t) - x(s)|^2)$, we see that $x(t)$ converges to $x(\infty)$ in L^2 as $t \uparrow \infty$. In particular, for A in $\mathscr{F}_T$,

$$E^P(x(\infty); A) = \lim_{s \to \infty} E^P(x(T + s); A) = \lim_{s \to \infty} E^P(x(T); A) = E^P(x(T); A).$$

Thus

$$x(T) = E^P(x(\infty)|\mathscr{F}_T) \quad \text{a.s. } P \quad \text{for all} \quad T \geq 0. \tag{1.9}$$

This completes our discussion of the martingale convergence theorem. We now turn to the converse of 3.1.9.

3.1.23. Exercise. Let $\mathscr{M}$ be the collection of all sets A in $\mathscr{F}$ that are independent of a given collection $\mathscr{N}$, i.e.,

$$P(A \cap B) = P(A)P(B)$$

for all B in $\mathscr{N}$. Show that $\mathscr{M}$ is a σ-algebra.

3.1.24. Exercise. Let $\mathscr{M}$ be the σ-algebra of all sets A in $\mathscr{F}$ independent of $\sigma(\eta(t), t \geq 0)$, where $\eta(\cdot)$ is an $(\Omega, \mathscr{F}_t, P)$ Brownian motion and $\mathscr{F}$ is any σ-algebra containing all the $\mathscr{F}_t$'s. For each $t \geq 0$, let $\bar{\mathscr{F}}_t, t \geq 0$, be the smallest σ-algebra containing $\mathscr{F}_t$ and $\mathscr{M}$. Show that $\eta(\cdot)$ is an $(\Omega, \bar{\mathscr{F}}_t, P)$ Brownian motion. Note in particular that $\mathscr{M}$ contains all sets of measure zero.

This shows that we may always assume that a random variable θ independent of $\eta(\cdot)$ is $\mathscr{F}_0$-measurable, without loss of generality, since $\mathscr{F}_t$ may be replaced by $\bar{\mathscr{F}}_t$ and θ is then $\bar{\mathscr{F}}_0$-measurable.

3.1.25. Exercise. For $a \geq 0$ set $\log^+ a = \max(\log a, 0)$. Show that for $a, b \geq 0$ one has $b \log^+ a \leq (1/e)a + b \log^+ b$.

3.1.26. Exercise. Use 3.1.25 to conclude that for f, g as in 3.1.19 one has

$$E^P(f) \leq \frac{e}{e - 1}(1 + E^P(g \log^+ g)).$$

3.1.27. Exercise. This is an L^1 version of Doob's L^r martingale inequality 3.1.20, $r > 1$. Let $x(\cdot)$ be an $(\Omega, \mathscr{F}_t, P)$ martingale. Show that

$$E^P\left(\sup_{0 \leq t \leq T} |x(t)|\right) \leq \frac{e}{e - 1}(1 + E^P(\varphi(|x(T)|)))$$

for all $T \geq 0$, where $\varphi(a) = a \log^+ a$.

In what follows we shall frequently use stopping times to stop progressively measurable processes. To this end the following is necessary.

3.1.28. Exercise. Let $x(\cdot)$ be progressively measurable and right continuous and let τ be a stopping time. Show that $x_\tau(\cdot)$ is progressively measurable and right continuous, where $x_\tau(t, \omega) = x(t \wedge \tau(\omega), \omega)$.

3.2. Stochastic Calculus

In this section we review the basic results of stochastic calculus; these center around the stochastic integral due to K. Ito (1944).

Let $(\Omega, \mathscr{F}, P)$ be a probability space and let $\mathscr{F}_t$, $t \geq 0$, be a nondecreasing family of sub-σ-algebras of $\mathscr{F}$. Let $\eta(\cdot)$ be an $\mathbb{R}^m$-valued $(\Omega, \mathscr{F}_t, P)$ Brownian motion. Let $g(\cdot)$ be a progressively measurable (n by m matrix)-valued process satisfying

$$P\left(\int_0^T |g(t)|^2 \, dt < \infty, \ T > 0\right) = 1. \tag{2.1}$$

Then there is a progressively measurable right continuous P-almost surely continuous process valued in $\mathbb{R}^n$,

$$x(T) = \int_0^T g(t) \, d\eta(t) \colon \Omega \to \mathbb{R}^n, \qquad T \geq 0, \tag{2.2}$$

the *stochastic integral* of $g(\cdot)$, determined almost surely P. We discuss some of its properties and its construction.

Linearity is a property that should hold for any concept of integral; (2.2) is no exception:

$$P\left(C_1 \int_0^T g_1(t) \, d\eta(t) + C_2 \int_0^T g_2(t) \, d\eta(t)\right.$$
$$\left. = \int_0^T (C_1 g_1(t) + C_2 g_2(t)) \, d\eta(t), \ T \geq 0\right) = 1.$$

Here $g_1(\cdot)$, $g_2(\cdot)$ satisfy (2.1) while C_1, C_2 are p by n matrices.

In case $g(\cdot)$ satisfies the *square integrability* condition

$$E\left(\int_0^T |g(t)|^2 \, dt\right) < \infty, \qquad T > 0, \tag{2.3}$$

then one has

$$E^P\left(\left|\int_0^T g(t) \, d\eta(t)\right|^2\right) = E^P\left(\int_0^T |g(t)|^2 \, dt\right). \tag{2.4}$$

Moreover, in this case (2.2) is an $(\Omega, \mathscr{F}_t, P)$ martingale,

$$E^P\left(\int_0^t g(r) \, d\eta(r) | \mathscr{F}_s\right) = \int_0^s g(r) \, d\eta(r) \quad \text{a.s. } P$$

for all $t > s \geq 0$; in particular, the *expectation* of (2.2) is then *zero*.

We recall the steps leading to the construction of (2.2). First, a process $g(\cdot)$ is *simple* if $g(\cdot)$ is progressively measurable, bounded ($|g(t, \omega)| \leq$ constant for all t and ω), and for some $n \geq 1$ satisfies $g(t) = g([nt]/n)$, $t \geq 0$. In this case (2.2) is defined by

$$x(T) = \int_0^T g(t)\, d\eta(t) = \sum_{k=0}^{\infty} g\left(\frac{k}{n}\right)\left(\eta\left(T \wedge \left(\frac{k+1}{n}\right)\right) - \eta\left(T \wedge \frac{k}{n}\right)\right). \quad (2.5)$$

Using 3.1.12 and 3.1.13, one can then check that (2.5) satisfies (2.4) and is a progressively measurable right continuous P-almost surely continuous martingale. Hence, by Doob's inequality (1.5) and (2.4), $x(\cdot)$ satisfies

$$E^P\left(\sup_{0 \leq t \leq T} \left|x(t) - \int_0^t e(s)\, d\eta(s)\right|^2\right) \leq 4E^P\left(\int_0^T |g(t) - e(t)|^2\, dt\right) \quad (2.6)$$

for *all* simple $e(\cdot)$ and simple $g(\cdot)$. Now for any $g(\cdot)$ satisfying (2.3) there is a sequence of simple processes $g_n(\cdot)$, $n \geq 1$, satisfying

$$\int_0^T |g(t) - g_n(t)|^2\, dt \to 0 \qquad \text{as} \quad n \uparrow \infty \quad (2.7)$$

in the mean for all $T > 0$; moreover, $g_n(\cdot)$ can be chosen to satisfy

$$P\left(\int_0^{\infty} |g_n(t)|^2\, dt \leq \int_0^{\infty} |g(t)|^2\, dt\right) = 1$$

for all $n \geq 1$. Let $x_k(\cdot)$ denote the stochastic integral of $g_k(\cdot)$; then by Doob's inequality (1.4) and (2.4)

$$P\left(\sup_{0 \leq t \leq T} |x_j(t) - x_k(t)| \geq \varepsilon\right) \to 0 \qquad \text{as} \quad j, k \uparrow \infty \quad \text{for all} \quad \varepsilon > 0. \quad (2.8)$$

Thus there is a progressively measurable right continuous P-almost surely continuous process $x(\cdot)$ satisfying

$$\lim_{j \to \infty} P\left(\sup_{0 \leq t \leq T} |x_j(t) - x(t)| \geq \varepsilon\right) = 0 \qquad \text{for all} \quad \varepsilon > 0. \quad (2.9)$$

By passing to a subsequence, we can assume that $g_j(\cdot)$ is a sequence satisfying (2.7) in the mean and

$$P\left(\lim_{j \to \infty} \sup_{k \geq j} \sup_{0 \leq t \leq T} |x_k(t) - x_j(t)| = 0\right) = 1.$$

3.2.1. Exercise. Which subsequence?

Thus $x_j(t, \omega) \to x(t, \omega)$ uniformly on $0 \leq t \leq T$ for P-almost all ω. By Doob's inequality (1.5)

$$\lim_{j \to \infty} E^P\left(\sup_{0 \leq t \leq T} |x(t) - x_j(t)|^2\right) \leq \lim_{j \to \infty} \liminf_{k \to \infty} E^P\left(\sup_{0 \leq t \leq T} |x_k(t) - x_j(t)|^2\right)$$

$$\leq 4 \lim_{j\to\infty} \liminf_{k\to\infty} E^P\left(\int_0^T |g_k(t) - g_j(t)|^2 \, dt\right) \quad (2.10)$$

$$= 4 \lim_{j\to\infty} E^P\left(\int_0^T |g(t) - g_j(t)|^2 \, dt\right)$$

$$= 0.$$

3.2.2. Exercise. Use (2.10) to conclude that there is a P-almost surely unique process $x(\cdot)$ that is progressively measurable, right continuous, P-almost surely continuous, and satisfies (2.6) for all $T \geq 0$ and all simple $e(\cdot)$.

The process $x(\cdot)$ is the stochastic integral of $g(\cdot)$ against $\eta(\cdot)$ in the square-integrable case.

3.2.3. Exercise. Use (2.10) to conclude that the stochastic integral of $g(\cdot)$ satisfies (2.4) and is an $(\Omega, \mathscr{F}_t, P)$ martingale.

3.2.4. Exercise. Let $g(\cdot)$ satisfy (2.3) and let τ be a stopping time. Show that

$$P\left(\int_0^{T\wedge\tau} g(t) \, d\eta(t) = \int_0^T 1_{t<\tau} g(t) \, d\eta(t), \, T \geq 0\right) = 1.$$

Let $f(\cdot)$ be an $\mathbb{R}^n$-valued progressively measurable process satisfying

$$P\left(\int_0^T |f(t)| \, dt < \infty, \, T > 0\right) = 1. \quad (2.11)$$

3.2.5. Exercise. Show that there is a P-almost surely unique progressively measurable right continuous P-almost surely continuous process $x(\cdot)$ such that $P(x(t) = \int_0^t f(s) \, ds, \, t \geq 0) = 1$.

We are now ready to define the stochastic integral of $g(\cdot)$ in the general case (2.1). By 3.2.5, there exists a progressively measurable right continuous P-almost surely continuous process $I(\cdot)$ such that $I(t, \omega)$ equals the integral of $|g(\cdot, \omega)|^2$ over $[0, t]$, $t \geq 0$, for almost all ω in Ω. Let τ_n be the contact time of $I(\cdot)$ with the closed set $C_n = \{x \mid |x| \geq n\}$ (Section 3.1). Since $I(\cdot)$ is P-almost surely continuous and (2.1) holds, the contact time and the entrance time agree almost surely; thus

$$\tau_n = \inf\left\{T > 0 \,\middle|\, \int_0^T |g(t)|^2 \, dt \geq n\right\}$$

almost surely. In particular, $P(\tau_n \uparrow \infty \text{ as } n \uparrow \infty) = 1$ and $g_n(t) = 1_{t<\tau} g(t), \, t \geq 0$, $\tau = \tau_n$, satisfies

$$P\left(\int_0^\infty |g_n(t)|^2 \, dt \leq n\right) = 1$$

for all $n \geq 1$. Now, by 3.2.4, the stochastic integrals $x_n(\cdot)$ of $g_n(\cdot)$ are consistently defined in the sense of 3.1.15. Thus there exists a P-almost surely unique progressively measurable right continuous P-almost surely continuous process $x(\cdot)$ satisfying $P(x(t) = x_n(t), 0 \leq t < \tau_n) = 1$ for all $n \geq 1$. This process $x(\cdot)$ is the stochastic integral (2.2) of $g(\cdot)$ in general.

We now turn to the *Ito differential rule*. Let $f(\cdot)$ be an $\mathbb{R}^n$-valued progressively measurable process satisfying (2.11) and let $g(\cdot)$ be an n by m matrix-valued progressively measurable process satisfying (2.1). Let $x(\cdot)$ be an $\mathbb{R}^n$-valued progressively measurable right continuous P-almost surely continuous process satisfying

$$P\left(x(T) = x(0) + \int_0^T f(t)\, dt + \int_0^T g(t)\, d\eta(t), \ T \geq 0\right) = 1. \qquad (2.12)$$

Let $\varphi: \mathbb{R}^n \to \mathbb{R}$ be a C^2 function and set

$$\left(\frac{\partial \varphi}{\partial x} f\right)(t, \omega) = \sum_{i=1}^n \frac{\partial \varphi}{\partial x_i}(x(t, \omega)) f_i(t, \omega),$$

$$\left(\frac{\partial \varphi}{\partial x} g\right)(t, \omega)_j = \sum_{i=1}^n \frac{\partial \varphi}{\partial x_i}(x(t, \omega)) g_{ij}(t, \omega),$$

$$\frac{\partial^2 \varphi}{\partial x^2}(t, \omega)_{ij} = \frac{\partial^2 \varphi}{\partial x_i \partial x_j}(x(t, \omega)),$$

$$(L\varphi)(t, \omega) = \left(\frac{\partial \varphi}{\partial x} f\right)(t, \omega) + \tfrac{1}{2}\, \text{trace}\left(g^* \frac{\partial^2 \varphi}{\partial x^2} g\right)(t, \omega).$$

Then $L\varphi(\cdot)$ satisfies (2.11), $(\partial \varphi / \partial x) g(\cdot)$ satisfies (2.1), and

$$P\left(\varphi(x(T)) = \varphi(x(0)) + \int_0^T (L\varphi)(t)\, dt + \int_0^T \left(\frac{\partial \varphi}{\partial x} g\right)(t)\, d\eta(t), \ T \geq 0\right) = 1. \qquad (2.13)$$

This is the Ito differential rule.

3.2.6. Exercise. Let $z(\cdot)$ be an $\mathbb{R}^m$-valued progressively measurable process satisfying (2.1) and set

$$R(T) = \exp\left(-\int_0^T z(t)^*\, d\eta(t) - \tfrac{1}{2}\int_0^T |z(t)|^2\, dt\right), \qquad T \geq 0. \qquad (2.14)$$

Show that $R(\cdot)$ satisfies

$$P\left(R(T) = 1 - \int_0^T R(t) z(t)^*\, d\eta(t), \ T \geq 0\right) = 1. \qquad (2.15)$$

3.2.7. Lemma. *Let $z(\cdot)$ and $R(\cdot) = R(\cdot\,; z)$ be as in 3.2.6 with*

$$P\left(\int_0^\infty |z(t)|^2\, dt \leq c\right) = 1$$

*for some c > 0. Then R(·) is an (Ω, ℱ_t, P) martingale with $E^P(R(T)^2)$ bounded
as a function of T ≥ 0.*

PROOF. Let $C_n = \{x||x| \geq n\}$ and let τ_n be the contact time of R(·) with C_n.
Since R(·) is P-almost surely continuous, we have $P(\tau_n = \inf\{T > 0|R(T) \geq n\}) = 1$ and $P(\tau_n \uparrow \infty$ as $n \uparrow \infty) = 1$. By 3.1.28, 3.2.4, and (2.15), $R(\cdot \wedge \tau_n)$ is
a martingale. Noting that $R(\cdot \wedge \tau_n; 2z) \leq n^2$ almost surely, and applying
what we just learnt, we see that $R(\cdot \wedge \tau_n; 2z)$ is also a martingale so
$E^P(R(t \wedge \tau_n; 2z)) = E^P(R(0; 2z)) = 1$. Since $R(t \wedge \tau_n)^2 \leq R(t \wedge \tau_n; 2z)e^c$ almost surely follows that

$$E^P(R(T)^2) \leq \sup_n E^P(R(T \wedge \tau_n)^2) \leq e^c \qquad \text{for all} \quad T > 0.$$

This yields

$$\limsup_n E^P(|R(T) - R(T \wedge \tau_n)|)$$

$$\leq \limsup_n E^P(|R(T)1_{R(T) \leq a} - R(T \wedge \tau_n)1_{R(T \wedge \tau) \leq a}|) \qquad (\tau = \tau_n)$$

$$+ \sup_n E^P(R(T); R(T) \geq a)$$

$$+ \sup_n E^P(R(T \wedge \tau_n); R(T \wedge \tau_n) \geq a)$$

$$\leq 0 + \frac{1}{a}e^c + \frac{1}{a}e^c \to 0 \qquad \text{as} \quad a \uparrow \infty.$$

Thus $R(T \wedge \tau_n) \to R(T)$ in L^1 as $n \uparrow \infty$ which is enough to conclude that R(·)
is an (Ω, ℱ_t, P) martingale. □

In particular, 3.2.7 implies that $E^P(R(T)) = 1$ for all T > 0. Also the martingale convergence theorem applies and thus $R(\infty)$ exists and, moreover,
satisfies $E^P(R(\infty)) = 1$. R(·) is *the exponential martingale.*

Let $f: \mathbb{R}^n \to \mathbb{R}^n$ and $g: \mathbb{R}^n \to \mathbb{R}^{n \times m}$ be vector- and matrix-valued functions
of x, and assume that f and g satisfy a Lipschitz condition

$$|f(x) - f(x')| + |g(x) - g(x')| \leq \text{constant} \times |x - x'|. \qquad (2.16)$$

Let $\xi: \Omega \to \mathbb{R}^n$ be an $\mathscr{F}_0$-measurable random variable satisfying $E^P(|\xi|^2) < \infty$.
Then there is a progressively measurable right continuous P-almost surely
continuous process x(·) satisfying

$$P\left(x(T) = \xi + \int_0^T f(x(t))\, dt + \int_0^T g(x(t))\, d\eta(t), T \geq 0\right) = 1. \qquad (2.17)$$

Moreover, any two such processes agree in the sense that for any stopping
time τ one has $P(x(t) = x'(t), 0 \leq t < \tau) = 1$ whenever x'(·) satisfies

$$P\left(x'(T) = \xi + \int_0^T f(x'(t))\, dt + \int_0^T g(x'(t))\, d\eta(t), 0 \leq T < \tau\right) = 1. \qquad (2.18)$$

Equation (2.17) is *a stochastic differential equation*. Note that, by 3.1.16, $x(\cdot)$ can be chosen to be progressively measurable relative to $\sigma[\eta(s), 0 \le s \le t]$, $t \ge 0$.

The solution $x(\cdot)$ of (2.17) is obtained as the limit of *successive approximations*: let $x_0(\cdot) = \xi$ and let

$$x_{k+1}(T) = \xi + \int_0^T f(x_k(t)) \, dt + \int_0^T g(x_k(t)) \, d\eta(t), \qquad T \ge 0.$$

By means of estimates similar to (2.6), one can then show that there is a progressively measurable right continuous P-almost surely continuous process $x(\cdot)$ such that

$$\sup_{0 \le t \le T} |x_k(t) - x(t)|^2 \to 0 \qquad \text{as} \quad k \uparrow \infty$$

in the mean for all $T > 0$. This defines the solution of (2.17) in case (2.16) holds.

In general, f and g are only C^1 in x and no growth condition is assumed. In this case, there is a stopping time ζ and a progressively measurable right continuous P-almost surely continuous process $x(\cdot)$ satisfying

$$P\left(x(T) = \xi + \int_0^T f(x(t)) \, dt + \int_0^T g(x(t)) \, d\eta(t), 0 \le T < \zeta \right) = 1,$$

and

$$\lim_{T \uparrow \zeta} |x(T)| = \infty \qquad \text{on } \{\zeta < \infty\} \quad \text{a.s. } P. \tag{2.19}$$

Moreover, $(x(\cdot), \zeta)$ are uniquely determined by the above in the sense that for any stopping time τ and progressively measurable right continuous P-almost surely continuous $x'(\cdot)$ satisfying (2.18) one has $P(\tau \le \zeta) = 1$ and $P(x(t) = x'(t), 0 \le t < \tau) = 1$. ζ is *the explosion time* of (2.17).

3.2.8. Exercise. Suppose $g_n(\cdot)$, $n \ge 1$, satisfy (2.1) and

$$\int_0^T |g_n(t)|^2 \, dt \to 0 \qquad \text{as} \quad n \uparrow \infty$$

in probability for all $T > 0$. Show that

$$\sup_{0 \le t \le T} \left| \int_0^t g_n(s) \, d\eta(s) \right| \to 0 \qquad \text{as} \quad n \uparrow \infty$$

in probability for all $T > 0$.

3.3. Cameron–Martin–Girsanov Formula

Let $(\Omega, \mathscr{F}, P)$ be a probability space and let $\mathscr{F}_t$ be a nondecreasing family of sub-σ-algebras of $\mathscr{F}$. Let $\eta(\cdot)$ be an $(\Omega, \mathscr{F}_t, P)$ Brownian motion, valued in $\mathbb{R}^m$, and let $z(\cdot)$ be an $\mathbb{R}^m$-valued progressively measurable process satisfying

$$P\left(\int_0^T |z(t)|^2 \, dt < \infty, \, T \geq 0 \right) = 1. \tag{3.1}$$

Let $y(\cdot)$ be an $\mathbb{R}^m$-valued progressively measurable right continuous P-almost surely continuous process satisfying

$$P\left(y(t) = \int_0^t z(s) \, ds + \eta(t), \, t \geq 0 \right) = 1. \tag{3.2}$$

Set $\mathscr{Y}_T = \sigma[y(t), 0 \leq t \leq T]$, $\mathscr{Y}_\infty = \sigma[y(t), t \geq 0]$; let $R(\cdot)$ be given by (2.14). We begin by assuming that

$$P\left(\int_0^\infty |z(t)|^2 \, dt \leq c \right) = 1 \tag{3.3}$$

for some $c > 0$. Then, by Lemma 3.2.7, $R(\cdot)$ is a martingale with $E^P(R(T)^2)$ bounded in T. Thus by the martingale convergence theorem (Section 3.1) there exists $R(\infty)$ in $L^2(\Omega, \mathscr{F}, P)$ such that $R(T) \to R(\infty)$ in L^2 as $T \uparrow \infty$ and $R(T) = E^P(R(\infty)|\mathscr{F}_T)$ a.s. P for all $T \geq 0$. In particular, $E^P(R(T)) = 1$ for all $0 \leq T \leq \infty$. Define a probability measure Q on $(\Omega, \mathscr{F})$ by setting

$$Q(A) = E^P(R(\infty); A), \qquad A \text{ in } \mathscr{F},$$

the expectation of $R(\infty)$ over the set A. Then for A in $\mathscr{F}_T$, $Q(A) = E^P(R(T); A)$. In general, i.e., when (3.3) does not hold, the measure Q need not exist. Nevertheless, the measures $Q = Q_c$, defined for each value of $c > 0$, are consistently defined in a certain sense. Because of this, we work with $Q = Q_c$ one at a time and then "splice" the results together using 3.1.15 (see 3.3.4 and 3.3.5 below).

3.3.1. Lemma. $y(\cdot)$ is an $(\Omega, \mathscr{F}_t, P)$ Brownian motion.

PROOF. Note first that $Q = P$ on $\mathscr{F}_0$ so that $Q(y(0) = 0) = 1$. Second, since $R(T) > 0$, Q and P are mutually absolutely continuous on $\mathscr{F}_T$ for each $T > 0$; this implies that $y(\cdot)$ is Q-almost surely continuous. We now apply 3.1.10; let c be any row vector and let A be in $\mathscr{F}_s$ with $s < t \leq T$. Let $c(t) = 1_{t \leq T} c$. Then

$$E^Q\left(\exp\left(cy(t) - \frac{|c|^2 t}{2} \right); A \right)$$

$$= E^P\left(R(t) \exp\left(cy(t) - \frac{|c|^2 t}{2} \right); A \right)$$

$$= E^P\left(\exp\left(\int_0^t (c(r) - z(r)^*) \, d\eta(r) - \tfrac{1}{2} \int_0^t |c(r) - z(r)^*|^2 \, dr \right); A \right)$$

$$= E^P\left(\exp\left(\int_0^s (c(r) - z(r)^*) \, d\eta(r) - \tfrac{1}{2} \int_0^s |c(r) - z(r)^*|^2 \, dr \right); A \right)$$

$$= E^P\left(R(s)\exp\left(cy(s) - \frac{|c|^2 s}{2} \right); A \right) \qquad \text{(by 3.2.7)}$$

$$= E^Q\left(\exp\left(cy(s) - \frac{|c|^2 s}{2} \right); A \right). \qquad \Box$$

This result will be useful in Chapter 4. Now Lemma 3.3.1 and 3.1.9 combined imply that $\mathcal{Y}_\infty$ and $\mathcal{F}_0$ are independent under Q. Since $y(\cdot)$ is an $(\Omega, \mathcal{F}_t, Q)$ Brownian motion, in particular $y(\cdot)$ is an $(\Omega, \mathcal{Y}_t, Q)$ Brownian motion (Exercise 3.1.16). To avoid confusion from now on "$\mathcal{F}_t$-progressively measurable" will be referred to simply as "progressively measurable" while progressive measurability relative to $\mathcal{Y}_t$ will always be specified by saying "$\mathcal{Y}_t$-progressively measurable."

Let $g(\cdot)$ be a $\mathcal{Y}_t$-progressively measurable process that satisfies (2.1). The rest of the section is devoted to defining the stochastic integral

$$\int_0^T g(t)\, dy(t), \quad T \geq 0,$$

as a $\mathcal{Y}_t$-progressively measurable right continuous P-almost surely continuous process and establishing some related facts. We begin under the assumption that (3.3) holds. Then the measure Q exists and, by Lemma 3.3.1 and the results of Section 3.2, the stochastic integral is well defined, $\mathcal{Y}_t$-progressively measurable, right continuous, Q-almost surely continuous, and is the Q-almost surely unique such process satisfying the Q-analogue of (2.6).

3.3.2. Lemma. *The following holds:*

$$P\left(\int_0^T g(t)\, dy(t) = \int_0^T g(t)z(t)\, dt + \int_0^T g(t)\, d\eta(t),\ T \geq 0 \right) = 1. \quad (3.4)$$

PROOF. Assume first that $P(\int_0^\infty |g(t)|^2\, dt \leq k) = 1$ for some $k > 0$. Choose simple $\mathcal{Y}_t$-progressively measurable processes $g_n(\cdot)$ approximating $g(\cdot)$ as in Section 3.2. Clearly, (3.4) holds for $g_n(\cdot)$. We need to show that we can take the limit.

Since the sequence (2.7) is uniformly bounded, by Exercise 3.1.8 (2.7) holds in the Q-mean as well. Thus by the Q-analogue to (2.6)

$$\int_0^T g_n(t)\, dy(t) \to \int_0^T g(t)\, dy(t)$$

in $L^2(\Omega, \mathcal{F}_T, Q)$ and hence in Q-probability and hence (Exercise 3.1.7) in P-probability. Also

$$\int_0^T g_n(t)\, d\eta(t) \to \int_0^T g(t)\, d\eta(t)$$

in $L^2(\Omega, \mathscr{F}_T, P)$ and hence in P-probability. Finally,

$$\int_0^T g_n(t)z(t)\,dt \to \int_0^T g(t)z(t)\,dt$$

in P-probability. Thus we can take the limit in (3.4). For the general case let τ_n be the contact time of $\int_0^\cdot |g(s)|^2\,ds$ with C_n as in Section 3.2. Then almost surely

$$\tau_n = \inf\left\{T > 0 \,\Big|\, \int_0^T |g(t)|^2\,dt \geq n\right\}.$$

Then, by what we have just seen, (3.4) holds for $g_n(\cdot)$ where $g_n(t) = g(t)1_{t<\tau}$, $\tau = \tau_n$. Appealing to 3.2.4 and noting that $P(\tau_n \uparrow \infty$ as $n \uparrow \infty) = 1$, the result follows.
$\square$

We note that since P and Q are mutually absolutely continuous on $\mathscr{F}_T$ for all $T > 0$, the Q-stochastic integral of $g(\cdot)$ against $y(\cdot)$ is also P-almost surely continuous. We shall need a variant of 3.3.2 in the context of stochastic differential equations.

3.3.3. Corollary. *Let x^0 be in $\mathbb{R}^n$. Let $f: \mathbb{R}^n \to \mathbb{R}^n$ and $g: \mathbb{R}^n \to \mathbb{R}^{n \times m}$ be C^1. Suppose that there is a progressively measurable right continuous P-almost surely continuous solution $x(\cdot)$ to*

$$P\left(x(T) = x^0 + \int_0^T (f(x(t)) + g(x(t))z(t))\,dt + \int_0^T g(x(t))\,d\eta(t),\ T \geq 0\right) = 1.$$
(3.5)

Then there is a $\mathscr{Y}_t$-progressively measurable right continuous P-almost surely continuous solution $\bar{x}(\cdot)$ to (3.5).

PROOF. First assume that the first partial derivatives of f, g are bounded. Then the solution $x(\cdot)$ exists and can be obtained as the limit of successive approximations: set $x_0(\cdot) = x^0$ and

$$x_{k+1}(T) = x^0 + \int_0^T (f(x_k(t)) + g(x_k(t))z(t))\,dt + \int_0^T g(x_k(t))\,d\eta(t), \quad T \geq 0.$$

By induction and 3.3.2, we can choose $x_k(\cdot)$ to be $\mathscr{Y}_t$-progressively measurable for all k. Since $x_k(\cdot)$ converges to $x(\cdot)$ in P-probability, there exists a $\mathscr{Y}_t$-progressively measurable limit $\bar{x}(\cdot)$. In particular $P(\bar{x}(t) = x(t), t \geq 0) = 1$ and so $\bar{x}(\cdot)$ satisfies (3.5).

Second, for each $n \geq 1$ choose f_n, g_n to equal f, g on $|x| \leq n$ and to have bounded first partial derivatives. Then by the above, for each n we can choose a $\mathscr{Y}_t$-progressively measurable solution $x_n(\cdot)$ to (3.5) with f_n, g_n replacing f, g. Let τ_n be the contact time of $x_n(\cdot)$ with C_n. Then τ_n is a $\mathscr{Y}_t$ stopping time and

$P(x_n(t) = x(t), 0 \leq t < \tau_n) = 1$. Thus

$$P(\tau_n \leq T) = P\left(\sup_{0 \leq t \leq \tau} |x_n(t)| = n, \tau_n \leq T \right)$$

$$= P\left(\sup_{0 \leq t \leq \tau} |x(t)| = n, \tau_n \leq T \right) \qquad (\tau = \tau_n)$$

$$\leq P\left(\sup_{0 \leq t \leq T} |x(t)| \geq n \right) \to 0 \qquad \text{as} \quad n \uparrow \infty.$$

Thus $P(\tau_n \uparrow \infty \text{ as } n \uparrow \infty) = 1$; applying 3.1.15 yields a $\mathcal{Y}_t$-progressively measurable right continuous P-almost surely continuous $\bar{x}(\cdot)$ satisfying $P(\bar{x}(t) = x_n(t), 0 \leq t < \tau_n) = 1$ for all $n \geq 1$. Thus $\bar{x}(\cdot)$ is the required solution. $\qquad \square$

We now turn to the question of what can be done when (3.3) is not assumed to hold. A natural assumption is then that there exists a sequence of stopping times τ_n increasing to ∞ and processes $z^n(\cdot)$, $y^n(\cdot)$, $n \geq 1$, such that

$$P\left(y^n(t) = \int_0^t z^n(s)\, ds + \eta(t), t \geq 0 \right) = 1, \tag{3.6}$$

$$P\left(\int_0^\infty |z^n(t)|^2\, dt \leq n \right) = 1, \tag{3.7}$$

and $z^n(\cdot)$, $y^n(\cdot)$ agree with $z(\cdot)$, $y(\cdot)$ up to time τ_n. More precisely, we say that (3.2) is *localizable* if for all $n \geq 1$ there is a progressively measurable right continuous P-almost surely continuous process $y^n(\cdot)$, a progressively measurable process $z^n(\cdot)$, a $\mathcal{Y}_t^n = \sigma(y^n(s), 0 \leq s \leq t)$-stopping time τ_n, and a $\mathcal{Y}_t$-stopping time σ_n such that (3.6) and (3.7) hold, and

$$P(\tau_n \uparrow \infty \text{ as } n \uparrow \infty) = 1,$$

$$y^n(t \wedge \sigma_n) = y(t \wedge \sigma_n) \qquad t \geq 0, \text{ identically,}$$

$$P(z^n(t) = z(t), 0 \leq t < \tau_n) = 1,$$

$$P(\tau_n = \sigma_n) = 1 \qquad \text{for all} \quad n \geq 1.$$

We shall see that the set-up described in Chapter 4 is localizable. We now show that the results of 3.3.2 and 3.3.3 can be extended to the localizable situation. Note that localizability implies (3.1).

3.3.4. Proposition. *Assume that (3.2) is localizable. Then there is a $\mathcal{Y}_t$-progressively measurable right continuous P-almost surely continuous process $\int_0^t g(s)\, dy(s), t \geq 0$, satisfying (3.4).*

PROOF. From Section 3.1, $g(\cdot)$ can be represented as

$$g(t) = F(t, y(t \wedge t_1), y(t \wedge t_2), \ldots)$$

for some sequence $t_k \geq 0$, $k \geq 1$. Set

$$g^n(t) = F(t, y^n(t \wedge t_1), y^n(t \wedge t_2), \ldots) 1_{t < \tau}, \qquad \tau = \tau_n.$$

Then $g^n(\cdot)$ is $\mathscr{Y}_t^n$-progressively measurable and

$$P\left(\int_0^T |g^n(t)|^2 \, dt \leq \int_0^T |g(t)|^2 \, dt \right) = 1$$

and thus $g^n(\cdot)$ satisfies (2.1). Let $I^n(\cdot)$ denote the stochastic integral of $g^n(\cdot)$ against $y^n(\cdot)$ as in 3.3.2. Representing $I^n(\cdot)$ in terms of $y^n(\cdot)$ explicitly as above, we can find a $\mathscr{Y}_t$-progressively measurable right continuous process $J^n(\cdot)$ such that $P(I^n(t \wedge \tau_n) = J^n(t \wedge \tau_n), t \geq 0) = 1$. Now by the fact $I^n(\cdot)$ and hence $J^n(\cdot)$ satisfy an equality similar to (3.4) for all n, we see that $J^n(\cdot)$ are consistently defined in the sense of 3.1.15. Thus there exists a $\mathscr{Y}_t$-progressively measurable right continuous P-almost surely continuous process satisfying (3.4). □

3.3.5. Exercise. Assuming localizability, prove 3.3.3.

3.3.6. Exercise. In the proof of 3.3.4, where did we use the fact that σ_n is a $\mathscr{Y}_t$-stopping time?

3.4. Notes and References

The exposition in this chapter is almost entirely from [3.8]. However, Section 3.1 is not a good place to get aquainted with Brownian motion for the first time. For a first-time look, see [3.7] and then [3.9].

We emphasize that the family of σ-algebras $\mathscr{F}_t$, $t \geq 0$, is *arbitrary*, i.e., not necessarily complete nor right continuous. Also here right continuity of the paths does not mean "cadlag." We do this because we wish the filtering result in Chapter 4 to hold with $\mathscr{Y}_t$, $t \geq 0$, uncompleted. This takes a little extra work, but we feel that the logical clarity gained more than offsets this. If we had completed our σ-algebras then the Completeness Lemma would be elementary. As stated, however, it is a deep and important fact, and forms the basis for the existence of "nice" versions. Note that if we assume anything less than right continuity for our sample paths, then the proof of 3.1.4 would not hold. On the other hand, if we assume anything more than right continuity, such as existence of left limits, then the proof of the Completeness Lemma would not hold.

The statement concerning the representability of progressively measurable functions is Exercise 1.5.6 of [3.8]. A good text for some of the basic probability theory is [3.1]. The Completeness Lemma is 4.3.3 of [3.8]. Property (i) of the martingale definition is never used in the text. The Wiener measure is first defined in [3.10]. Doob's theorems are in [3.2]. The stochastic integral is first defined in [3.4]. Approximation of integrands by simple ones appears in [3.8] as 4.3.2. The Ito differential rule is 4.4.1, 4.6.11 of [3.8]. Lemma 3.2.7 is a special case of a result in [3.5]. The theorems concerning the stochastic differential equations are in [3.6]. Another good source for results in this chapter is [3.3].

[3.1]   R. B. Ash, *Real Analysis and Probability*, Academic Press, New York, 1972.

[3.2]   J. L. Doob, *Stochastic Processes*, Wiley, New York, 1952.

[3.3]   N. Ikeda and S. Watanabe, *Stochastic Differential Equations and Diffusion Processes*, North-Holland, Amsterdam, 1981.

[3.4]   K. Ito, "Stochastic Integral," *Proc. Imperial Acad. Tokyo*, **20** (1944), 519–524.

[3.5]   R. S. Liptser and A. N. Shiryayev, *Statistics of Random Processes*, Vol. I and II, Springer-Verlag, New York, 1977.

[3.6]   H. P. McKean, *Stochastic Integrals*, Academic Press, New York, 1969.

[3.7]   E. Nelson, *Dynamical Theories of Brownian Motion*, Princeton University Press, Princeton, NJ, 1967.

[3.8]   D. W. Stroock and S. R. S. Varadhan, *Multidimensional Diffusion Processes*, Springer-Verlag, New York, 1979.

[3.9]   S. R. S. Varadhan, *Diffusion Problems and Partial Differential Equations*, Tata Lecture Notes, Bombay, 1980.

[3.10] N. Wiener, "Differential Space," *J. Math. Phys.*, **2** (1923), 131–174.

CHAPTER 4
Filtering

4.1. Filtering

Let $(\Omega, \mathcal{F}, P)$ be a probability space and let $\mathcal{F}_t$, $t \geq 0$, be a nondecreasing family of sub-σ-algebras of $\mathcal{F}$. Let $\eta(\cdot)$ be an $(\Omega, \mathcal{F}_t, P)$ Brownian motion in $\mathbb{R}^m$. Let $\theta: \Omega \to \{1, \ldots, N\}$ be $\mathcal{F}_0$-measurable with $P(\theta = j) = \pi_j^0$; here $\{\pi_1^0, \ldots, \pi_N^0\}$ is a fixed but arbitrary distribution on $\{1, \ldots, N\}$. For each $j = 1, \ldots, N$ let $z_j(\cdot)$ be a progressively measurable process. Throughout this chapter we shall assume that

$$P\left(\max_{1 \leq j \leq N} \int_0^T |z_j(t)|^2 \, dt < \infty, T > 0 \right) = 1. \tag{1.1}$$

Set $z(\cdot) = z_\theta(\cdot)$ and assume $y(\cdot)$ is an $\mathbb{R}^m$-valued progressively measurable right continuous P-almost surely continuous process satisfying

$$P\left(y(t) = \int_0^t z(s) \, ds + \eta(t), t \geq 0 \right) = 1. \tag{1.2}$$

Set $\mathcal{Y}_T = \sigma[y(t), 0 \leq t \leq T]$, $T \geq 0$, and $\mathcal{Y}_\infty = \sigma[y(t), t \geq 0]$.

Throughout this chapter we shall assume that $z_j(\cdot)$ is $\mathcal{Y}_t$-progressively measurable for all $j = 1, \ldots, N$.

Of interest in this chapter are the conditional probabilities $P(\theta = j | \mathcal{Y}_T)$, $j = 1, \ldots, N$. The motivation for studying the conditional probabilities when given (1.2) is straightforward: the process $\eta(\cdot)$ represents the *noise*, while the N values of θ represent N different *states* of the *system*, and $z_j(\cdot)$ represents the *signal* emitted when the system is in state j. Because of the presence of the noise $\eta(\cdot)$ in (1.2), one has access only to the *observations* $y(\cdot)$. The problem then is to determine the state given the observations $y(t), 0 \leq t \leq T$, up to time

T. In this context, the requirement that $z_j(\cdot)$ be $\mathscr{Y}_t$-progressively measurable makes precise the notion that the *j*th signal is known once the observation is known.

One is also interested in determining the dynamical behavior (time evolution) of $P(\theta = j|\mathscr{Y}_T)$, since in applications one needs to compute the conditional probability in *real time*; this means that $P(\theta = j|\mathscr{Y}_{T+\Delta T})$ should be computable solely from knowledge of $P(\theta = j|\mathscr{Y}_T)$ and the incoming observations $y(t)$, $T \le t \le T + \Delta T$. The precise interpretation of this turns out to be the fact that the conditional probability satisfies a stochastic differential equation which is determined in this section.

Another question of interest concerns the increase in the amount of *information* concerning θ contained in $y(t)$, $0 \le t \le T$, as *T* increases. A related question is the large time behavior of $P(\theta = j|\mathscr{Y}_T)$ as $T\uparrow\infty$, in particular whether or not one *learns* the "true" value of θ after observing $y(\cdot)$ for all time.

We begin by deriving an explicit formula for $P(\theta = j|\mathscr{Y}_T)$ that will serve us well in what follows. Heuristically, 4.1.2 below is simply Bayes' rule coupled with the Cameron–Martin–Girsanov result.

To apply the results of Section 3.3 we need to check that (1.2) is *localizable* in the sense defined there. More precisely, we need to establish the following proposition. Recall that unless specified by a qualifier, progressive measurability always refers to $\mathscr{F}_t$, $t \ge 0$.

4.1.1. Proposition. *For each $n \ge 1$, there is a progressively measurable right continuous P-almost surely continuous process $y^n(\cdot)$, a $\mathscr{Y}_t^n = \sigma(y^n(s), 0 \le s \le t)$-progressively measurable process $z_j^n(\cdot)$ for each $j = 1, \ldots, N$, a $\mathscr{Y}_t^n$-stopping time τ_n and a $\mathscr{Y}_t$-stopping time σ_n such that with $z^n(\cdot) = z_\theta^n(\cdot)$,*

(i) $P(\sigma_n\uparrow\infty \text{ as } n\uparrow\infty) = 1$,

(ii) $y(t \wedge \sigma_n) = y^n(t \wedge \sigma_n)$, $t \ge 0$, identically,

(iii) $P\left(\int_0^\infty |z^n(t)|^2 \, dt \le n\right) = 1$,

(iv) $P(\tau_n = \sigma_n) = 1$,

(v) $P(z_j^n(t) = z_j(t), 0 \le t < \tau_n) = 1$,

(vi) $P\left(y^n(t) = \int_0^t z^n(s) \, ds + \eta(t), t \ge 0\right) = 1$.

PROOF. By 3.2.5 there exists a $\mathscr{Y}_t$-progressively measurable right continuous *P*-almost surely continuous process $I_j(\cdot)$ such that for each $j = 1, \ldots, N$, $P(I_j(t) = \int_0^t |z_j(s)|^2 \, ds, t \ge 0) = 1$. Let σ_n be the contact time of $I(\cdot) = \max_{1 \le j \le N} I_j(\cdot)$ with the closed set $C_n = \{x||x| \ge n\}$. Then since $I(\cdot)$ is *P*-almost surely continuous, the contact time and the entrance time agree almost surely; thus using (1.1) we have

$$\sigma_n = \inf\left\{T > 0 \,\middle|\, \max_{1 \le j \le N} \int_0^T |z_j(t)|^2 \, dt \ge n\right\} \tag{1.3}$$

almost surely. Clearly then σ_n is a $\mathscr{Y}_t$-stopping time and (i) holds. For the remainder of the proof, $n \geq 1$ is fixed. Now for each $j = 1, \ldots, N$, represent $z_j(\cdot)$ as

$$z_j(t, \omega) = F_j(t, y(t \wedge t_{1j}, \omega), y(t \wedge t_{2j}, \omega), \ldots), \qquad t \geq 0, \quad \omega \text{ in } \Omega,$$

for some $t_{ij} \geq 0$ and measurable F_j (Section 3.1). Let $J(\cdot)$ be a progressively measurable right continuous version of $\int_0^{\cdot} z(s)\, ds$ and set

$$y^n(t, \omega) = y(t, \omega) - J(t, \omega) + J(t \wedge \sigma_n(\omega), \omega),$$

$$w_j^n(t, \omega) = F_j(t, y^n(t \wedge t_{1j}, \omega), y^n(t \wedge t_{2j}, \omega), \ldots), \qquad t \geq 0, \quad \omega \text{ in } \Omega.$$

Then (ii) holds and $w_j^n(\cdot)$ is $\mathscr{Y}_t^n$-progressively measurable. Set

$$z_j^n(t) = w_j^n(t) 1 \left(\int_0^t |w_j^n(s)|^2\, ds \leq n \right), \qquad t \geq 0.$$

Then (Section 3.1) $z_j^n(\cdot)$ is $\mathscr{Y}_t^n$-progressively measurable and

$$\max_{1 \leq j \leq N} \int_0^\infty |z_j^n(t)|^2\, dt \leq n$$

identically on Ω and so (iii) holds and

$$\tau_n = \inf \left\{ T > 0 \,\Big|\, \max_{1 \leq j \leq N} \int_0^T |z_j^n(t)|^2\, dt \geq n \right\} \tag{1.4}$$

is a $\mathscr{Y}_t^n$-stopping time. To establish (iv) first note that by (ii) and (1.3) $z_j^n(t) = w_j^n(t) = z_j(t)$, $0 \leq t < \sigma_n$, almost surely. Second, we claim that $\tau_n \leq \sigma_n$ almost surely: indeed by (1.3) again and the above ($\sigma = \sigma_n$)

$$\max_{1 \leq j \leq N} \int_0^\sigma |z_j^n(t)|^2\, dt = n \qquad \text{almost surely on } \{\sigma_n < \infty\}.$$

It follows then, by (1.4), that $\tau_n \leq \sigma_n$ almost surely. Third, this inequality and reversing the roles of σ_n and τ_n yields the reverse inequality. This establishes (iv). (v) and (vi) are then immediate. $\qquad \square$

In particular, (1.2) is localizable whenever (1.1) holds and so the results of Section 3.3 apply. Thus the stochastic integrals

$$\int_0^T z_j(t)^*\, dy(t), \qquad T \geq 0,$$

$j = 1, \ldots, N$, exist as $\mathscr{Y}_t$-progressively measurable right continuous, P-almost surely continuous processes satisfying

$$P\left(\int_0^T z_j(t)^*\, dy(t) = \int_0^T z_j(t)^* z(t)\, dt + \int_0^T z_j(t)^*\, d\eta(t), T \geq 0 \right) = 1,$$

$j = 1, \ldots, N$. Now set

$$l_j(T) = \exp\left(\int_0^T z_j(t)^* \, dy(t) - \tfrac{1}{2} \int_0^T |z_j(t)|^2 \, dt \right), \qquad T \geq 0,$$

$$R(T) = \exp\left(-\int_0^T z(t)^* \, d\eta(t) - \tfrac{1}{2} \int_0^T |z(t)|^2 \, dt \right), \qquad T \geq 0,$$

$$\pi_j(T) = \frac{l_j(T)\pi_j^0}{\sum_{k=1}^N l_k(T)\pi_k^0}, \qquad T \geq 0. \tag{1.5}$$

Then $R(\cdot)$, $l_j(\cdot)$, and $\pi_j(\cdot)$ are progressively measurable, right continuous, P-almost surely continuous, while $l_j(\cdot)$ and $\pi_j(\cdot)$ are in addition $\mathscr{Y}_t$-progressively measurable. Note that $\pi_j(0) = \pi_j^0, j = 1, \ldots, N$, and, by 3.3.4,

$$P(R(T)^{-1} = l_\theta(T), T \geq 0) = 1.$$

The following is fundamental to the chapter.

4.1.2. Theorem. *For all* $T \geq 0$, A *in* $\mathscr{Y}_T, j = 1, \ldots, N$,

$$P(\{\theta = j\} \cap A) = E^P(\pi_j(T); A); \tag{1.6}$$

thus

$$P(\theta = j | \mathscr{Y}_T) = \pi_j(T) \quad \text{a.s. } P \tag{1.7}$$

PROOF. First we assume that there is a $c > 0$ such that

$$P\left(\int_0^\infty |z(t)|^2 \, dt \leq c \right) = 1. \tag{1.8}$$

Then there is a probability measure Q on $(\Omega, \mathscr{F})$ satisfying $dQ/dP = R(T)$ on $\mathscr{F}_T$ for all $T \geq 0$ (Section 3.3). Moreover, $y(\cdot)$ is then an $(\Omega, \mathscr{F}_t, Q)$ Brownian motion (3.3.1) and so $y(\cdot)$ and θ are Q-independent (3.1.9). Thus

$$
\begin{aligned}
P(\{\theta = j\} \cap A) &= E^Q(R(T)^{-1}; \{\theta = j\} \cap A) \\
&= E^Q(l_j(T); \{\theta = j\} \cap A) \\
&= E^Q(l_j(T); A)Q(\theta = j) && (y(\cdot), \theta \ Q\text{-ind.}) \\
&= E^Q(l_j(T); A)\pi_j^0 && (Q = P \text{ on } \mathscr{F}_0) \\
&= \sum_{k=1}^N E^Q(l_k(T)\pi_j(T); A)\pi_k^0 && (\text{by (1.5)}) \\
&= \sum_{k=1}^N E^Q(l_k(T)\pi_j(T); A)Q(\theta = k) && (Q = P \text{ on } \mathscr{F}_0) \\
&= \sum_{k=1}^N E^Q(l_k(T)\pi_j(T); \{\theta = k\} \cap A) && (y(\cdot), \theta \ Q\text{-ind.})
\end{aligned}
$$

$$= \sum_{k=1}^{N} E^Q(R(T)^{-1}\pi_j(T); \{\theta = k\} \cap A)$$

$$= E^Q(R(T)^{-1}\pi_j(T); A)$$

$$= E^P(\pi_j(T); A).$$

Now we *drop* the assumption that $z(\cdot)$ satisfy (1.8). Let $z_j^n(\cdot)$, $y^n(\cdot)$, τ_n, and σ_n be as in 4.1.1. Let $l_j^n(\cdot)$, $\pi_j^n(\cdot)$ be defined analogously to $l_j(\cdot)$, $\pi_j(\cdot)$ but with $z_j^n(\cdot)$, $y^n(\cdot)$ replacing $z_j(\cdot)$, $y(\cdot)$. Then, by 3.2.4,

$$P(\pi_j^n(T) = \pi_j(T), 0 \le T < \tau_n) = 1.$$

Let A be in $\mathscr{Y}_T$. Then $a(\cdot) = 1_A 1_{[T, \infty)}$ is $\mathscr{Y}_t$-progressively measurable. Representing this explicitly as in Section 3.1, we can find a $\mathscr{Y}_t^n$-progressively measurable process $a^n(\cdot)$ such that $P(a(t) = a^n(t), 0 \le t < \tau_n) = 1$. Thus $1_A 1_{T < \tau}(\tau = \tau_n) = a(T)1_{T < \tau} = a^n(T)1_{T < \tau}$ is $\mathscr{Y}_T^n$-measurable and so $A \cap \{T < \tau_n\}$ is in $\mathscr{Y}_T^n$. Hence

$$P(\{\theta = j\} \cap A) = \lim_{n \uparrow \infty} P(\{\theta = j\} \cap A \cap \{T < \tau_n\})$$

$$= \lim_{n \uparrow \infty} E^P(\pi_j^n(T); A \cap \{T < \tau_n\})$$

$$= \lim_{n \uparrow \infty} E^P(\pi_j(T); A \cap \{T < \tau_n\})$$

$$= E^P(\pi_j(T); A). \qquad \square$$

We will need the limiting case $T \to \infty$ of 4.1.2. Note that $\mathscr{Y}_\infty$ is the smallest σ-algebra containing all the σ-algebras $\mathscr{Y}_t$, $t \ge 0$.

4.1.3. Theorem. *For* $j = 1, \ldots, N$,

$$\pi_j(\infty) = \lim_{T \uparrow \infty} \pi_j(T) \quad \text{exists a.s. } P \tag{1.9}$$

and

$$P(\theta = j | \mathscr{Y}_\infty) = \pi_j(\infty) \quad \text{a.s. } P. \tag{1.10}$$

PROOF. By (1.7), $\pi_j(T) = P(\theta = j | \mathscr{Y}_T)$ a.s. P, $T \ge 0$. Thus $\pi_j(\cdot)$ is a bounded $(\Omega, \mathscr{Y}_t, P)$ martingale. By Doob's martingale convergence theorem (Section 3.1) $\pi_j(\infty)$ exists and (1.9) holds. Now for A in $\mathscr{Y}_T$

$$E^P(\pi_j(\infty); A) = \lim_{s \uparrow \infty} E^P(\pi_j(T + s); A)$$

$$= \lim_{s \uparrow \infty} E^P(\pi_j(T); A)$$

$$= \lim_{s \uparrow \infty} P(\{\theta = j\} \cap A)$$

$$= P(\{\theta = j\} \cap A).$$

Now the collection of all sets A for which this equality holds is a σ-algebra and contains $\mathscr{Y}_T$ for all $T \ge 0$. Thus the equality holds for all A in $\mathscr{Y}_\infty$. This proves (1.10). $\qquad \square$

Note that we could not say that the process $P(\theta = j|\mathcal{Y}_T)$ is a martingale as it may not be right continuous. Indeed, (1.7) shows that by modifying this process on a null set, for each time, one obtains a martingale. Now set

$$\hat{z}(T) = \sum_{k=1}^{N} z_k(T)\pi_k(T)$$

$$= \sum_{k=1}^{N} z_k(T)P(\theta = k|\mathcal{Y}_T) \quad \text{a.s. } P. \tag{1.11}$$

If $z(T)$ is integrable, then this equals (Exercise 3.1.12)

$$\hat{z}(T) = \sum_{k=1}^{N} E^P(z_k(T)1_{\theta=k}|\mathcal{Y}_T) = E^P(z(T)|\mathcal{Y}_T) \quad \text{a.s. } P,$$

the conditional expectation of $z(T)$ given the observations $y(t)$, $0 \le t \le T$, up to time T. Note that $\hat{z}(\cdot)$ is $\mathcal{Y}_t$-progressively measurable and satisfies $P(\int_0^T |\hat{z}(t)|^2\, dt < \infty,\ T > 0) = 1$.

4.1.4. Theorem. *For $j = 1, \ldots, N$,*

$$P\left(\pi_j(t) = \pi_j^0 + \int_0^t \pi_j(s)(z_j(s) - \hat{z}(s))^*(dy(s) - \hat{z}(s)\, ds),\ t \ge 0\right) = 1. \tag{1.12}$$

PROOF. Apply the Ito rule (2.13), Chapter 3, to $\varphi(x) = e^x$, $x(t) = \log l_j(t)$ to yield

$$P\left(l_j(t) = 1 + \int_0^t l_j(s)z_j(s)^*\, dy(s),\ t \ge 0\right) = 1.$$

Apply the Ito rule again to $\varphi(x_1, \ldots, x_N) = x_j\pi_j^0/\sum_{k=1}^{N} x_k\pi_k^0$ and $x_j(t) = l_j(t)$, $t \ge 0$. The result follows. $\qquad\square$

4.1.5. Exercise. Verify the application of the Ito rule in 4.1.4.

In (1.12) "dy" can be interpreted as "$d\eta + z\, dt$" or as a Q-Brownian differential "dy" as in Section 3.3. It turns out that "$dy - \hat{z}\, dt$" can be interpreted as a P-Brownian differential "dv"; this has the effect of making (1.12) somewhat easier to handle.

The *innovations* process $v(\cdot)$ is

$$v(t) = y(t) - \int_0^t \hat{z}(s)\, ds, \quad t \ge 0.$$

Then $v(\cdot)$ is $\mathcal{Y}_t$-*progressively measurable*, right continuous, and P-almost surely continuous.

4.1.6. Theorem. $v(\cdot)$ *is an $(\Omega, \mathcal{Y}_t, P)$ Brownian motion.*

PROOF. By definition

$$P\left(v(t) = \int_0^t (z(s) - \hat{z}(s))\, ds + \eta(t),\ t \ge 0\right) = 1.$$

Apply the Ito rule to $\varphi(v(\cdot))$ where $\varphi(x) = e^{icx}$, $i = \sqrt{-1}$; here c is any row vector in $\mathbb{R}^m$. Then

$$P\left(e^{icv(t)} = 1 + i\int_0^t c(z(s) - \hat{z}(s))e^{icv(s)}\, ds + i\int_0^t e^{icv(s)}c\, d\eta(s)\right.$$

$$\left. - \tfrac{1}{2}|c|^2 \int_0^t e^{icv(s)}\, ds, \, t \geq 0\right) = 1.$$

Let $\sigma = \sigma_n$ be stopping times as in 4.1.1; then

$$P\left(e^{icv(t \wedge \sigma)} = e^{icv(s \wedge \sigma)} + i\int_s^t c(z(r) - \hat{z}(r))e^{icv(r)}1_{r<\sigma}\, dr\right.$$

$$\left. + i\int_s^t 1_{r<\sigma}e^{icv(r)}c\, d\eta(r) - \tfrac{1}{2}|c|^2 \int_s^t e^{icv(r)}1_{r<\sigma}\, dr\right) = 1, \qquad (1.13)$$

whenever $t \geq s \geq 0$. Now the conditional expectation $E^P(\cdot \,|\, \mathcal{Y}_s)$ of the stochastic integral in (1.13) is zero as its integrand is bounded. Moreover, the conditional expectation of the first integral in (1.13) is also zero: for A in $\mathcal{Y}_s$

$$E^P\left(i\int_s^t cz(r)e^{icv(r)}1_{r<\sigma}\, dr;\, A\right)$$

$$= i\int_s^t E^P(cz(r)e^{icv(r)}1_{r<\sigma};\, A)\, dr$$

$$= \sum_{k=1}^N i\int_s^t E^P(cz_k(r)1_{\theta=k}e^{icv(r)}1_{r<\sigma};\, A)\, dr$$

$$= \sum_{k=1}^N i\int_s^t E^P(cz_k(r)\pi_k(r)e^{icv(r)}1_{r<\sigma};\, A)\, dr \qquad \text{(because } A \text{ in } \mathcal{Y}_r)$$

$$= i\int_s^t E^P(c\hat{z}(r)e^{icv(r)}1_{r<\sigma};\, A)\, dr$$

$$= E^P\left(i\int_s^t c\hat{z}(r)e^{icv(r)}1_{r<\sigma}\, dr;\, A\right).$$

Thus $j(t) = E^P(e^{icv(t)};\, A)$ satisfies

$$j(t) = \lim_{n\uparrow\infty} E^P(e^{icv(t \wedge \sigma)};\, A)$$

$$= \lim_{n\uparrow\infty} E^P(e^{icv(s \wedge \sigma)};\, A) - \tfrac{1}{2}|c|^2 \int_s^t E^P(1_{r<\sigma}e^{icv(r)};\, A)\, dr$$

$$= j(s) - \tfrac{1}{2}|c|^2 \int_s^t j(r)\, dr, \qquad t \geq s.$$

Thus $j(\cdot)$ is continuous hence differentiable; thus $\dot{j} = -\tfrac{1}{2}|c|^2 j$. Solving this ordinary differential equation, we have

$$E^P(e^{icv(t)}; A) = \exp(-\tfrac{1}{2}|c|^2(t-s))E^P(e^{icv(s)}; A). \qquad (1.14)$$

Now by definition (Section 3.1) all we need verify is

$$P(v(t) \in B; A) = \int_B E^P(g_m(t-s, x-v(s)); A) \, dx$$

for all A in $\mathscr{Y}_s$ and B in $\mathscr{B}(\mathbb{R}^m)$. But (1.14) implies

$$E^P(e^{icv(t)}; A) = \int_{\mathbb{R}^m} e^{icx} E^P(g_m(t-s, x-v(s)); A) \, dx.$$

Thus, in order to obtain the result, we need to invert the Fourier transform in this last equation. This is a standard result which we now derive. Let φ be an integrable function on $\mathbb{R}^m$ and let φ_ε denote the convolution of φ with $g_m(\varepsilon, \cdot)$, $\varepsilon > 0$. Then the Fourier transform $\hat{\varphi}_\varepsilon$ of φ_ε is proportional to $\hat{\varphi}\hat{g}_m(\varepsilon, c) = \hat{\varphi} \exp(-\tfrac{1}{2}\varepsilon|c|^2)$ and is therefore also integrable. Thus

$$\varphi_\varepsilon(x) = \int_{\mathbb{R}^m} \hat{\varphi}_\varepsilon(c)e^{icx} \, dc.$$

Hence

$$\begin{aligned}
E^P(\varphi_\varepsilon(v(t)); A) &= \int_{\mathbb{R}^m} \hat{\varphi}_\varepsilon(c)E^P(e^{icv(t)}; A) \, dc \\
&= \int_{\mathbb{R}^m} \hat{\varphi}_\varepsilon(c) \int_{\mathbb{R}^m} e^{icx} E^P(g_m(t-s, x-v(s)); A) \, dx \, dc \\
&= \int_{\mathbb{R}^m} \left(\int_{\mathbb{R}^m} \hat{\varphi}_\varepsilon(c)e^{icx} \, dc \right) E^P(g_m(t-s, x-v(s)); A) \, dx \\
&= \int_{\mathbb{R}^m} \varphi_\varepsilon(x)E^P(g_m(t-s, x-v(s)); A) \, dx.
\end{aligned}$$

Now if φ is also continuous and bounded, then φ_ε converges to φ boundedly as $\varepsilon \downarrow 0$. Thus letting $\varepsilon \downarrow 0$ it follows that

$$E^P(\varphi(v(t)); A) = \int_{\mathbb{R}^m} \varphi(x)E^P(g_m(t-s, x-v(s)); A) \, dx.$$

Letting φ converge to 1_B, the result follows. $\qquad\square$

We want to show that $\pi_j(\cdot)$ satisfies the following variant of (1.12):

$$P\left(\pi_j(T) = \pi_j^0 + \int_0^T \pi_j(t)(z_j(t) - \hat{z}(t))^* \, dv(t), \ T \ge 0 \right) = 1. \qquad (1.12)$$

Here the stochastic integral is well defined by 4.1.6.

More generally, the innovations process can be defined to be any $\mathcal{Y}_t$-progressively measurable right continuous P-almost surely continuous process $v(\cdot)$ satisfying

$$P\left(y(t) = \int_0^t \hat{z}(s)\, ds + v(t),\, t \geq 0 \right) = 1. \tag{1.15}$$

Then, by examining the proof of 4.1.6, one sees that the result still holds.

4.1.7. Proposition. *For each $j = 1, \ldots, N$, (1.12) holds. Note that this explicitly exhibits $\pi_j(\cdot)$ as a stochastic integral.*

PROOF. We want to apply 3.3.4 with $v(\cdot)$ replacing $\eta(\cdot)$ and $\mathcal{Y}_t$ replacing $\mathcal{F}_t$. To do so we have to check that (1.15) is localizable. But this follows from 4.1.1 by taking $N = 1$, replacing $z_1(\cdot)$ by $\hat{z}(\cdot)$, $\eta(\cdot)$ by $v(\cdot)$, and $\mathcal{F}_t$ by $\mathcal{Y}_t$ in the statement of 4.1.1. Thus 3.3.4 applies and so

$$P\left(l_j(t) = \exp\left(\int_0^t z_j(s)^*\, dv(s) + \int_0^t z_j(s)^* \hat{z}(s)\, ds \right.\right.$$
$$\left.\left. - \tfrac{1}{2} \int_0^t |z_j(s)|^2\, ds \right),\, t \geq 0 \right) = 1.$$

Now a straightforward computation using the Ito rule yields the result.   □

If we let $\mathcal{N}_t = \sigma(v(s), 0 \leq s \leq t)$ then 4.1.6 implies that $\mathcal{N}_t \subset \mathcal{Y}_t$ for all $t \geq 0$. An interesting and difficult question, the *innovations problem*, concerns the validity of the reverse inclusion. We show here that the reverse inclusion holds in a special case, provided one ignores sets of measure zero.

4.1.8. Theorem. *Suppose that $z_j(\cdot)$, $j = 1, \ldots, N$, are all $\mathcal{N}_t$-progressively measurable. Then $y(\cdot)$ is almost surely equal to an $\mathcal{N}_t$-progressively measurable process and $\bar{\mathcal{N}}_t = \bar{\mathcal{Y}}_t$ for all $t \geq 0$. Here $\bar{\mathcal{M}}$ denotes P-completion of the sub-σ-algebra $\mathcal{M}$ in $\mathcal{F}$.*

PROOF. The second statement is an immediate consequence of the first. The result is an immediate consequence of the fact (3.3.5) that (1.12) is a stochastic differential equation driven by $v(\cdot)$ and so must have an $\mathcal{N}_t$-progressively measurable solution and (1.15).   □

4.1.9. Exercise. Show that

$$P\left(\sum_{k=1}^N l_k(t)\pi_k^0 = \exp\left(\int_0^t \hat{z}(s)^*\, dy(s) - \tfrac{1}{2} \int_0^t |\hat{z}(s)|^2\, ds \right),\, t \geq 0 \right) = 1.$$

(*Hint:* Apply the Ito rule to the quotient of the left-hand side over the right-hand side.)

In particular, note that 4.1.8 holds in the special "independent signal plus noise" case of $z_j(\cdot)$ depending only on time.

4.2. Consistency

Let $(\Omega, \mathscr{F}, P)$ be a probability space and let $\mathscr{F}_t$, $t \geq 0$, be a nondecreasing family of sub-σ-algebras of $\mathscr{F}$. We assume we are in the situation established at the beginning of Section 4.1. Clearly then, (1.1) implies

$$P\left(\int_0^T |z(t)|^2 \, dt < \infty, T > 0\right) = 1. \tag{2.1}$$

Given $\varphi: \{1, \ldots, N\} \to \mathbb{R}$ (i.e., $\varphi = (\varphi_1, \ldots, \varphi_N)$ is in $\mathbb{R}^N$) let

$$\hat{\phi}(T) = \sum_{k=1}^N \varphi_k \pi_k(T) = E^P(\varphi(\theta)|\mathscr{Y}_T) \quad \text{a.s. } P, \tag{2.2}$$

where $\pi_j(\cdot)$ is given by (1.5). Then the process $\hat{\phi}(\cdot)$ is an $(\Omega, \mathscr{F}_t, P)$ martingale and so

$$\hat{\phi}(\infty) = \lim_{T \uparrow \infty} \hat{\phi}(T) = \sum_{k=1}^N \varphi_k \pi_k(\infty) \quad \text{exists a.s. } P,$$

by 4.1.3. Recall that when $\mathscr{M}$ is a sub-σ-algebra of $\mathscr{F}$, $\bar{\mathscr{M}}$ denotes the P-completion of $\mathscr{M}$ in $\mathscr{F}$, the σ-algebra generated by $\mathscr{M}$ and the null sets in $\mathscr{F}$.

4.2.1. Proposition. *The following are equivalent:*

(a) $P(\lim_{T \uparrow \infty} \hat{\phi}(T) = \varphi(\theta) \text{ for all } \varphi) = 1.$
(b) $P(\pi_\theta(\infty) = 1) = 1.$
(c) θ is $\bar{\mathscr{Y}}_\infty$-*measurable.*

4.2.2. Exercise. Prove 4.2.1.

If any of the equivalent statements in 4.2.1 hold, then we say that we have *consistency*. Thus consistency simply means that by observing $y(t)$, $0 \leq t \leq T$, for all time $T \geq 0$, one can recover the state θ. We caution that whether or not consistency holds depends on the signals $z_j(\cdot)$, $j = 1, \ldots, N$.

The purpose of this section is to derive a usable criterion for consistency. To do this we will need a preliminary estimate.

Recall *Holder's inequality*: for any nonnegative random variables f, g on $(\Omega, \mathscr{F}, P)$,

$$E^P(fg) \leq E^P(f^r)^{1/r} \cdot E^P(g^{r'})^{1/r'} \qquad \left(\frac{1}{r} + \frac{1}{r'} = 1\right)$$

for all $r > 1$. Also set

$$R_\alpha(T; z) = \exp\left(-\int_0^T z(t)^* \, d\eta(t) - \frac{\alpha}{2}\int_0^T |z(t)|^2 \, dt\right),$$

and note that $R_\alpha(T; z)^\alpha = R_1(T; \alpha z)$. Recall (by 3.2.7) that for any $z(\cdot)$ satisfying (1.8), $R_1(\cdot; z)$ is an $(\Omega, \mathscr{F}_t, P)$ martingale with $E^P(R_1(T; z)) = 1$ for $T \geq 0$. Here is the preliminary estimate.

4.2.3. Lemma. *Let $z(\cdot)$ be any progressively measurable process satisfying (2.1); let $s \geq 0$. Then*

$$P\left(\sup_{t \geq s} \frac{-\int_0^t z(u)^* \, d\eta(u)}{\int_0^t |z(u)|^2 \, du} \geq \varepsilon\right) \leq E^P\left(\exp\left(-\tfrac{3}{4}\varepsilon^2 \int_0^s |z(u)|^2 \, du\right)\right)^{1/3} \quad (2.3)$$

for all $\varepsilon > 0$.

PROOF. First notice that, by replacing z by εz, it is enough to prove the case $\varepsilon = 1$. Second, let us temporarily assume that (1.8) holds. Then

$$P\left(\sup_{s \leq t \leq T} \frac{-\int_0^t z(u)^* \, d\eta(u)}{\int_0^t |z(u)|^2 \, du} \geq 1\right)$$

$$= P\left(\sup_{s \leq t \leq T} \frac{\left(-\int_0^t z(u)^* \, d\eta(u) - \frac{1}{2}\int_0^t |z(u)|^2 \, du\right)}{\int_0^t |z(u)|^2 \, du} \geq \tfrac{1}{2}\right)$$

$$\leq P\left(\sup_{s \leq t \leq T} R_1(t; z) \geq \exp\left(\tfrac{1}{2}\int_0^s |z(u)|^2 \, du\right)\right).$$

Now let $\bar{R}(t) = R_1(t + s)\exp(-\frac{1}{2}\int_0^s |z(u)|^2 \, du)$, $\bar{\mathscr{F}}_t = \mathscr{F}_{t+s}$, $t \geq 0$. Then $\bar{R}(\cdot)$ is an $(\Omega, \bar{\mathscr{F}}_t, P)$ martingale. Applying (1.4) of Chapter 3, the above is

$$= P\left(\sup_{0 \leq t \leq T-s} \bar{R}(t) \geq 1\right)$$

$$\leq E^P\left(R_1(T; z)\exp\left(-\tfrac{1}{2}\int_0^s |z(u)|^2 \, du\right)\right)$$

$$= E^P\left(R_1(s; z)\exp\left(-\tfrac{1}{2}\int_0^s |z(u)|^2 \, du\right)\right)$$

$$= E^P\left(R_{3/2}(s; z)\exp\left(-\tfrac{1}{4}\int_0^s |z(u)|^2 \, du\right)\right)$$

$$\leq E^P \left(R_{3/2}(s; z)^{3/2} \right)^{2/3} \cdot E^P \left(\exp \left(-\tfrac{3}{4} \int_0^s |z(u)|^2 \, du \right) \right)^{1/3}$$

$$= E^P \left(R_1(s; \tfrac{3}{2}z) \right)^{2/3} \cdot E^P \left(\exp \left(-\tfrac{3}{4} \int_0^s |z(u)|^2 \, du \right) \right)^{1/3}$$

$$= E^P \left(\exp \left(-\tfrac{3}{4} \int_0^s |z(u)|^2 \, du \right) \right)^{1/3},$$

where we have used Holder's inequality with $r = \tfrac{3}{2}$, $r' = 3$. Letting $T \uparrow \infty$, we obtain (2.3) in the bounded case.

We now drop the assumption that (1.8) holds. Let $\tau = \tau_n$ be the contact time of the right continuous progressively measurable version of $\int_0 |z(t)|^2 \, dt$ with the set $\{x \, | \, |x| \geq n\}$. Then $P(\tau_n \uparrow \infty \text{ as } n \uparrow \infty) = 1$ and we can apply (2.3) to $z_n(\cdot)$, where $z_n(t) = 1_{t < \tau} z(t)$, $t \geq 0$, $\tau = \tau_n$. Thus, by 3.2.4,

$$P \left(\sup_{s \leq t < \tau_n} \frac{- \int_0^t z(u)^* \, d\eta(u)}{\int_0^t |z(u)|^2 \, du} \geq 1 \right)$$

$$\leq E^P \left(\exp \left(-\tfrac{3}{4} \int_0^{s \wedge \tau_n} |z(u)|^2 \, du \right) \right)^{1/3}.$$

Now, letting $n \uparrow \infty$, the result follows. □

4.2.4. Corollary. *Let $z(\cdot)$ be any progressively measurable process satisfying* (2.1) *and*

$$P \left(\int_0^\infty |z(t)|^2 \, dt = +\infty \right) = 1. \tag{2.4}$$

Then

$$P \left(\lim_{T \uparrow \infty} \frac{\int_0^T z(t)^* \, d\eta(t)}{\int_0^T |z(t)|^2 \, dt} = 0 \right) = 1. \tag{2.5}$$

PROOF. Applying (2.3) to $\pm z(\cdot)$ and letting $s \uparrow \infty$, we have

$$P \left(\limsup_{s \uparrow \infty} \frac{\left| \int_0^t z(t)^* \, d\eta(u) \right|}{\int_0^t |z(u)|^2 \, du} \geq \varepsilon \right)$$

$$\leq 2 E^P \left(\exp \left(-\tfrac{3}{4} \varepsilon^2 \int_0^\infty |z(u)|^2 \, du \right) \right)^{1/3} = 0$$

for all $\varepsilon > 0$. The result follows. □

4.2.5. Corollary. *Let A be in $\mathscr{F}_0$ and let $z(\cdot)$ be any progressively measurable process satisfying (2.1) and*

$$\int_0^\infty |z(t)|^2 \, dt = +\infty \quad \text{on } A \text{ a.s. } P.$$

Then

$$\lim_{T \uparrow \infty} \frac{\displaystyle\int_0^T z(t)^* \, d\eta(t)}{\displaystyle\int_0^T |z(t)|^2 \, dt} = 0 \quad \text{on } A \text{ a.s. } P.$$

PROOF. Set $\bar{z}(t, \omega) = z(t, \omega)$ on A, $t \geq 0$, and $\bar{z}(t, \omega) = 1$ off A, $t \geq 0$. Then $\bar{z}(\cdot)$ satisfies (2.4). Since

$$\int_0^T \bar{z}(t)^* \, d\eta(t) = \int_0^T z(t)^* \, d\eta(t), \qquad T \geq 0, \quad \text{on } A \text{ a.s. } P$$

(why?), applying (2.5), the corollary follows. $\qquad\square$

Let us now return to the situation described at the beginning of Section 4.1, with $z(t) = z_\theta(t)$, $t \geq 0$.

4.2.6. Theorem. *Consistency holds if and only if*

$$\int_0^\infty |z_j(t) - z(t)|^2 \, dt = +\infty \quad \text{on } \{\theta \neq j\} \text{ a.s. } P \tag{2.6}$$

for all j with $\pi_j^0 > 0$.

PROOF. Let j be such that $\pi_j^0 > 0$. Then on $\{\theta \neq k\}$, $L_k(T) \to 0$ as $T \uparrow \infty$ a.s. P by 4.2.5, where

$$L_k(T) = \exp\left(\int_0^T (z_k(t) - z(t))^* \, d\eta(t) - \tfrac{1}{2} \int_0^T |z_k(t) - z(t)|^2 \, dt \right).$$

Now (1.5) and elementary manipulation show that on $\{\theta = j\}$,

$$\pi_\theta(T)^{-1} = \frac{\sum_{k=1}^N L_k(T)\pi_k^0}{\pi_j^0} = 1 + \frac{\sum_{k \neq j} L_k(T)\pi_k^0}{\pi_j^0} \tag{2.7}$$

$$\to 1 \quad \text{as } T \uparrow \infty \quad \text{a.s. } P.$$

Conversely, consistency implies

$$\frac{\pi_j(T)}{\pi_\theta(T)} \to 0 \quad \text{as } T \uparrow \infty \quad \text{on } \{\theta \neq j\} \text{ a.s. } P.$$

When this is written out, this implies (2.6). $\qquad\square$

4.2.7. Exercise. Show that for any progressively measurable $z(\cdot)$ satisfying (2.1),

$$P\left(\sup_{T \geq 0} \exp\left(-\int_0^T z(t)^* \, d\eta(t) - \tfrac{1}{2}\int_0^T |z(t)|^2 \, dt\right) < \infty\right) = 1.$$

4.2.8. Exercise. Show that one always has

$$P(\pi_\theta(\infty) > 0) = 1$$

whether or not consistency holds. (*Hint*: Use (2.7).)

The following helps underscore the importance of consistency.

4.2.9. Exercise. Let $T \geq 0$ be *finite*. Show that θ is *not* $\overline{\mathcal{Y}}_T$-measurable.

4.2.10. Exercise. Suppose that there is a probability measure Q on $(\Omega, \mathcal{F})$ with respect to which $y(\cdot)$ is an $(\Omega, \mathcal{F}_t, Q)$ Brownian motion and $Q = P$ on $\mathcal{F}_0$. Show that the only time that θ is $\mathcal{Y}_\infty$-measurable is when θ is P-almost surely constant.

4.2.11. Exercise. Continuing 4.2.10, suppose that $Q \ll P$. Show that consistency does not hold except in the trivial case of θ being P-almost surely constant.

The last two exercises provide a good example of the fact that events of zero probability are not to be taken lightly in the present (infinite dimensional) context. Indeed, although we do not show this here, the existence of a measure Q as in 4.2.10 occurs frequently; however, the absolute continuity of Q relative to P does not always follow. Thus it is possible for θ to be $\overline{\mathcal{Y}}_\infty$-measurable but not $\mathcal{Y}_\infty$-measurable; this shows there is an appreciable difference between $\mathcal{Y}_\infty$ and $\overline{\mathcal{Y}}_\infty$, even from the point of view of applications.

The above paragraph provides one reason why one may not want to assume that the basic σ-algebras $\mathcal{F}_t$, $t \geq 0$, are complete. In other situations there are other reasons.

4.3. Shannon Information

Let $(\Omega, \mathcal{F}, P)$ be a probability space and let $\mathcal{F}_t$, $t \geq 0$, be a nondecreasing family of sub-σ-algebras of $\mathcal{F}$. Let $\eta(\cdot)$ be an $(\Omega, \mathcal{F}_t, P)$ Brownian motion in $\mathbb{R}^m$ and let $\theta: \Omega \to \{1, \ldots, N\}$ be $\mathcal{F}_0$-measurable and distributed according to $\pi^0 = \{\pi_1^0, \ldots, \pi_N^0\}$. Let $y(\cdot)$ be an $\mathbb{R}^m$-valued progressively measurable right continuous P-almost surely continuous process and let $z_j(\cdot)$, $j = 1, \ldots, N$, be $\mathcal{Y}_t$-progressively measurable processes satisfying (1.1). Set $z(\cdot) = z_\theta(\cdot)$ and suppose that

$$P\left(y(t) = \int_0^t z(s)\, ds + \eta(t), t \geq 0 \right) = 1. \tag{3.1}$$

We recall that the conditional probabilities satisfy

$$P(\theta = j | \mathcal{Y}_T) = \pi_j(T) \quad \text{a.s. } P, \qquad j = 1, \ldots, N, \quad 0 \leq T \leq \infty,$$

where $\pi_j(\cdot)$ is given by (1.5). The distribution $\pi(T) = \{\pi_1(T), \ldots, \pi_N(T)\}$ is then a measure of the amount of information concerning θ contained in $y(t)$, $0 \leq t \leq T$. In this section we seek a *numerical* measure of this information. We shall see that the Shannon information $I(T)$ of the pair θ and $y(t)$, $0 \leq t \leq T$, can be naturally described in terms of the data of the problem (3.1), and provides such a numerical measure.

We begin with the *information functional* of two measures. Let $(X, \mathcal{B})$ be an event space and let μ, v be probability measures on $\mathcal{B}$. Suppose, first, that μ is absolutely continuous with respect to v; set $R = d\mu/dv \geq 0$. Then the *information of μ relative to v* is the quantity

$$I(\mu; v) = E^v(R \log R) = E^\mu(\log R). \tag{3.2}$$

It turns out that $I \geq 0$ always; to see this we need *Jensen's inequality*:

For any nonnegative random variable f on a probability space $(\Omega, \mathcal{F}, P)$ and any *convex* $\varphi: [0, \infty) \to \mathbb{R}$,

$$\varphi(E^P(f)) \leq E^P(\varphi(f)).$$

In particular, $\varphi(\lambda) = \lambda \log \lambda$, $\lambda \geq 0$, is convex since $\varphi''(\lambda) = 1/\lambda > 0$. Thus

$$I(\mu; v) = E^v(\varphi(R)) \geq \varphi(E^v(R)) = \varphi(1) = 0.$$

In general, if μ is not absolutely continuous with respect to v, we set $I(\mu; v) = +\infty$.

A special case of (3.2) is when $X = \{1, \ldots, N\}$; then

$$I(\pi; \pi^0) = \sum_{k=1}^{N} \pi_k \log\left(\frac{\pi_k}{\pi_k^0}\right). \tag{3.3}$$

Another special case of (3.2) is when the measures μ, v relate in a certain way to two given random variables; this yields the Shannon information of the pair of random variables, defined as follows.

Let x_j be X_j-valued random variables on $(\Omega, \mathcal{F}, P)$, $j = 1, 2$, and let $\mathcal{B}_j$ denote the σ-algebra on X_j, $j = 1, 2$. Set

$$\mu_j(B) = P(x_j \in B), \qquad B \in \mathcal{B}_j,$$

and

$$\mu(B_1 \times B_2) = P(x_1 \in B_1, x_2 \in B_2).$$

Then μ_j is the distribution of x_j on X_j, $j = 1, 2$, and μ is the distribution of

(x_1, x_2) on $X_1 \times X_2$. Let $\mu_1 \times \mu_2$ denote the product of μ_1 and μ_2. The *Shannon information* of the pair x_1 and x_2 is defined to be

$$I(x_1, x_2) = I(\mu; \mu_1 \times \mu_2).$$

Suppose now that μ is absolutely continuous with respect to $\mu_1 \times \mu_2$ and let $R = d\mu/d(\mu_1 \times \mu_2) \geq 0$. Then R is the $(\mu_1 \times \mu_2)$-almost surely unique function on $X_1 \times X_2$ satisfying

$$E^P(f(x_1, x_2)) = E^\mu(f) = E^{\mu_1 \times \mu_2}(Rf).$$

By *Fubini's* theorem, this equals

$$
\begin{aligned}
E^P(f(x_1, x_2)) &= \int_{X_1} E^{\mu_2}(R(\lambda, \cdot)f(\lambda, \cdot))\, d\mu_1(\lambda) \\
&= \int_\Omega E^P(R(x_1(\omega), x_2)f(x_1(\omega), x_2))\, dP(\omega).
\end{aligned}
\tag{3.4}
$$

Also

$$I(x_1, x_2) = E^\mu(\log R) = E^P(\log R(x_1, x_2)). \tag{3.5}$$

Note that $I(x_1, x_2) = I(x_2, x_1)$ is symmetric.

The main result in this section is an explicit formula for the Shannon information of the pair $\theta, y(t), 0 \leq t \leq T$, where $y(\cdot)$ is the observation process (3.1).

Let $D([0, T]; \mathbb{R}^m)$ denote the set of all right continuous paths in $\mathbb{R}^m$ with σ-algebra $\mathcal{M}_T$ generated by the coordinate maps $\alpha \to \alpha(t), 0 \leq t \leq T$. Since $y(\cdot)$ is right continuous, it induces a map

$$y_T : \Omega \to D([0, T]; \mathbb{R}^m)$$

given by $(y_T(\omega))(t) = y(t, \omega)$. Similarly, we can define the event space $(D([0, \infty); \mathbb{R}^m), \mathcal{M}_\infty)$ and a naturally induced map $y_\infty : \Omega \to D([0, \infty); \mathbb{R}^m)$.

4.3.1. Exercise. For $0 \leq T \leq +\infty$, show that $y_T^{-1}(\mathcal{M}_T) = \mathcal{Y}_T$.

Thus for each $0 \leq T \leq +\infty$, θ and y_T are a pair of random variables on $(\Omega, \mathcal{F}, P)$ valued in $\{1, \ldots, N\}$ and $D([0, T]; \mathbb{R}^m)$ respectively $(D([0, \infty); \mathbb{R}^m)$ in case $T = +\infty)$. *We denote the Shannon information of this pair by* $I(T)$, $0 \leq T \leq +\infty$.

4.3.2. Proposition. *For* $0 \leq T \leq +\infty$,

$$I(T) = E^P(I(\pi(T); \pi^0)) = E^P(I(P(\theta = \cdot | \mathcal{Y}_T); \pi^0))$$

with $I(\pi; \pi^0)$ *given by* (3.3).

PROOF. Let $x_1 = \theta$ and $x_2 = y_T$; we will use (1.6), (3.4), and (3.5). Let A be in $\mathcal{Y}_T$; then, by 4.3.1, there is a B in $\mathcal{M}_T$ such that $1_A(\omega) = 1_B(y_T(\omega))$ for all ω in

Ω. Now

$$P(\{\theta = j\} \cap A) = E^P(1_{\theta=j} \cdot 1_B(y_T))$$

$$= \int_\Omega 1_{\theta(\omega)=j} E^P(R(\theta(\omega), y_T) 1_B(y_T)) \, dP(\omega) \qquad \text{(by (3.4))}$$

$$= E^P(R(j, y_T); A)\pi_j^0.$$

Thus (this is *Bayes' rule* for this context)

$$\pi_j(T) = R(j, y_T)\pi_j^0 \quad \text{a.s. } P,$$

which implies

$$I(T) = E^P(\log R(\theta, y_T))$$

$$= \sum_{k=1}^N E^P(1_{\theta=k} \log R(k, y_T))$$

$$= \sum_{k=1}^N E^P\left(1_{\theta=k} \log\left(\frac{\pi_k(T)}{\pi_k^0}\right)\right)$$

$$= \sum_{k=1}^N E^P\left(\pi_k(T) \log\left(\frac{\pi_k(T)}{\pi_k^0}\right)\right)$$

$$= E^P(I(\pi(T); \pi^0))$$

$$= E^P(I(P(\theta = \cdot | \mathscr{Y}_T); \pi^0)),$$

the last equality by (1.6). $\qquad\square$

4.3.3. Exercise. Show that $\log(N) \geq -\sum_{k=1}^N \pi_k \log \pi_k \geq 0$ for all probability distributions $\pi = \{\pi_1, \ldots, \pi_N\}$.

4.3.4. Corollary. *For all* $0 \leq T \leq +\infty$,

$$0 \leq I(T) \leq -\sum_{k=1}^N \pi_k^0 \log \pi_k^0.$$

PROOF. We can assume that $\pi_j^0 > 0$ for all j. Then, by 4.3.3,

$$0 \leq I(T) = E^P(I(\pi(T); \pi^0))$$

$$= \sum_{k=1}^N E^P(\pi_k(T) \log (\pi_k(T))) - \sum_{k=1}^N E^P(\pi_k(T) \log \pi_k^0)$$

$$\leq -\sum_{k=1}^N E^P(\pi_k(T) \log \pi_k^0)$$

$$= -\sum_{k=1}^N \pi_k^0 \log \pi_k^0. \qquad\square$$

We can now state the main result of this section.

4.3.5. Theorem. *For all* $0 \leq T \leq +\infty$,

$$I(T) = \tfrac{1}{2}E^P\left(\int_0^T |z(t) - \hat{z}(t)|^2 \, dt\right).$$

PROOF. Again both sides of the equation remain unchanged when we assume that $\pi_j^0 > 0$ for all j. For each $n \geq 1$, let $\varphi: \mathbb{R} \to \mathbb{R}$ be a C^∞ function such that $\varphi(a) = a$ for $a \geq 1/n$ and inf $\varphi > 0$ on $\mathbb{R}$. Define $H: \mathbb{R}^N \to \mathbb{R}$ by setting

$$H(\pi) = \sum_{j=1}^N \varphi(\pi_j) \log\left(\frac{\varphi(\pi_j)}{\varphi(\pi_j^0)}\right).$$

Then H is C^2 on $\mathbb{R}^N$. Let $\tau = \tau_n$ be a $\mathcal{Y}_t$-stopping time such that almost surely

$$\tau_n = \inf\left\{T > 0 \,\Big|\, \max_{1 \leq j \leq N} \int_0^T |z_j(t)|^2 \, dt \geq n\right\},$$

let $\sigma = \sigma_n$ be a $\mathcal{Y}_t$-stopping time such that almost surely

$$\sigma_n = \inf\left\{T > 0 \,\Big|\, \min_{1 \leq j \leq N} \pi_j(T) \leq \frac{1}{n}\right\},$$

and set $\rho_n = \tau_n \wedge \sigma_n$. Then $P(\rho_n \uparrow \infty \text{ as } n \uparrow \infty) = 1$. Now for $t < \rho = \rho_n$ $H(\pi(t)) = I(\pi(t); \pi^0)$ and

$$\frac{\partial^2 H}{\partial \pi_i \partial \pi_j}(\pi(t)) = \frac{\delta_{ij}}{\pi_i(t)}, \qquad i, j = 1, \dots, N. \tag{3.6}$$

Apply the Ito differential rule (2.13) of Chapter 3) to H and $\pi(\cdot)$ via (1.12). One obtains

$$P\left(H(\pi(T \wedge \rho)) = H(\pi^0) + \int_0^T 1_{t < \rho} \sum \frac{\partial H}{\partial \pi_k} \pi_k(z_k - \hat{z})^* \, dv \right.$$
$$\left. + \tfrac{1}{2}\int_0^{T \wedge \rho} \sum \pi_k |z_k - \hat{z}|^2 \, dt, \, T \geq 0\right) = 1, \tag{3.7}$$

using (3.6). Since the mean of the stochastic integral is zero, we have

$$E^P(I(\pi(T \wedge \rho); \pi^0)) = \tfrac{1}{2}E^P\left(\int_0^{T \wedge \rho} \sum \pi_k |z_k - \hat{z}|^2 \, dt\right)$$

$$= \tfrac{1}{2}E^P\left(\int_0^{T \wedge \rho} \sum 1_{\theta = k} |z_k - \hat{z}|^2 \, dt\right)$$

$$= \tfrac{1}{2}E^P\left(\int_0^{T \wedge \rho} |z - \hat{z}|^2 \, dt\right).$$

Letting $n \uparrow \infty$, the result follows *for T finite*. For $T = +\infty$, the result follows by letting $T \uparrow \infty$ and (1.9). $\qquad \square$

In particular, 4.3.5 implies that for any $\mathcal{Y}_t$-progressively measurable processes $z_j(\cdot), j = 1, \ldots, N$, satisfying (1.1) one has

$$E^P\left(\int_0^\infty |z(t) - \hat{z}(t)|^2 \, dt\right) < +\infty,$$

a fact which is not at all obvious.

We note that the formula 4.3.2 for $I(T)$ is not symmetrical in θ and $y(\cdot)$ (as opposed to (3.5)) and suggests that we think of $I(T)$ as *the information about θ contained in $y(t)$, $0 \le t \le T$.* Note also that 4.3.5 lends itself to the following appealing interpretation: *the rate of incoming information* concerning the state *is proportional to the averaged conditional variance of the signal.*

4.3.6. Exercise. Show that *consistency holds if and only if the Shannon information $I(\infty)$ of θ and $y(t)$, $0 \le t < \infty$, is as large as possible,* i.e., equal to $-\sum_{k=1}^N \pi_k^0 \log \pi_k^0$ (*Hint*: See 4.3.4.)

4.4. Notes and References

The results appearing in this chapter are all special cases of well-known results. The Bayes' formula (1.7) is a special case of a general result appearing in [4.6]. The dynamics of $\pi_j(\cdot)$ (equation (1.12)) is a special case of a general result appearing in [4.5] as well as in [3.5]. Theorem 4.1.8 is a special case of a general result [4.1]. Equation (1.12) also appears in [4.3]. The proof of Theorem 4.1.6 mimics the proof of Levy's theorem appearing in [4.4]. Theorem 4.3.5 is a special case of a result appearing in [4.2]. Although many of the results in the literature assume that the basic family of σ-algebras $\mathcal{F}_t$, $t \ge 0$, is "complete" and "right-continuous," we have made no such assumptions here (see Section 3.4).

Although the results described here assume that θ takes on finitely many values, we have endeavored to present them in such a way that they are extendible to the case when θ takes values in a complete metric space Λ. Taking Λ to be a path space then covers a wide class of filtering problems.

The "Kalman filter" and Gaussian "LQG" theory are not covered here as we include only the material necessary for Chapter 5. However, the central position the Kalman filter occupies cannot be overemphasized. A good survey is [4.7].

[4.1] D. F. Allinger and S. K. Mitter, "New Results on the Innovations Problem for Nonlinear Filtering," *Stochastics*, **4** (1981), 339–348.

[4.2] T. E. Duncan, "On the Calculation of Mutual Information," *SIAM J. Appl. Math.*, **19** (1970), 215–220.

[4.3] K. P. Dunn, "Measure Transformation, Estimation, Detection, and Stochastic Control," Ph.D. Dissertation, Washington University, St. Louis, MO, May 1974.

[4.4] R. Durrett, *Brownian Motion and Martingales in Analysis*, Wadsworth, Belmont, CA, 1984.

[4.5]   M. Fujisaki, G. Kallianpur, and H. Kunita, "Stochastic Differential Equations for the Nonlinear Filtering Problem," *Osaka J. Math.*, **9** (1972), 19–40.

[4.6]   G. Kallianpur and C. Streibel, "Estimation of Stochastic Processes," *Ann. Math. Statist.*, **39** (1968), 785–801.

[4.7]   J. C. Willems, "Recursive Filtering," *Statist. Neerlandica*, **32** (1978), 1–38.

CHAPTER 5
The Adaptive *LQ* Regulator

5.1. Introduction

Let $(\Omega, \mathscr{F}, P)$ be a probability space and let $\mathscr{F}_t$, $t \geq 0$, be a nondecreasing family of sub-σ-algebras of $\mathscr{F}$. Let $\eta(\cdot)$ be an $(\Omega, \mathscr{F}_t, P)$ Brownian motion valued in $\mathbb{R}^p$. For each distribution $\pi = \{\pi_1, \ldots, \pi_N\}$ on $\{1, \ldots, N\}$, let $\theta = \theta(\cdot; \pi): \Omega \to \{1, \ldots, N\}$ be an $\mathscr{F}_0$-measurable random variable distributed according to π.

Throughout a *control* refers to any progressively measurable $\mathbb{R}^m$-valued process $u(\cdot)$ satisfying

$$P\left(\int_0^T |u(t)|^2 \, dt < \infty, T > 0\right) = 1.$$

For each $j = 1, \ldots, N$ let n_j be a positive integer and let (A_j, B_j, C_j) be a triple with m inputs, n_j states, and p outputs. Let x_j^0 be an initial state in $\mathbb{R}^{n_j}$ for each $j = 1, \ldots, N$. Fix a distribution π^0 and set $\theta = \theta(\cdot; \pi^0)$.

We say that a control $u(\cdot)$ is *admissible* if there are progressively measurable right continuous almost surely continuous processes $x^u(\cdot)$, $y^u(\cdot)$ satisfying

$$P\left(x^u(t) = x_\theta^0 + \int_0^t [A_\theta x^u(s) + B_\theta u(s)] \, ds, t \geq 0\right) = 1, \qquad (1.1)$$

$$P\left(y^u(t) = \int_0^t C_\theta x^u(s) \, ds + \eta(t), t \geq 0\right) = 1, \qquad (1.2)$$

in such a way that $u(\cdot)$ is $\mathscr{Y}_t^u$-progressively measurable, where $\mathscr{Y}_t^u = \sigma[y^u(s), 0 \leq s \leq t]$. Now for each $j = 1, \ldots, N$, let $x_j^u(\cdot)$ denote the P-almost surely unique progressively measurable right continuous P-almost surely continuous solution to

$$P\left(x_j^u(t) = x_j^0 + \int_0^t [A_j x_j^u(s) + B_j u(s)]\, ds,\, t \geq 0\right) = 1.$$

By 3.2.5, $x_j^u(T)$ can be assumed to be $\sigma[u(t), 0 \leq t \leq T]$-measurable for all $T \geq 0$. Clearly, one then has

$$P(x^u(t) = x_\theta^u(t),\, t \geq 0) = 1.$$

Thus when $u(\cdot)$ is admissible $z_j^u(\cdot) = C_j x_j^u(\cdot)$ is $\mathscr{Y}_t^u$-progressively measurable, right continuous, and almost surely continuous, and $y^u(\cdot)$ satisfies

$$P\left(y^u(t) = \int_0^t z^u(s)\, ds + \eta(t),\, t \geq 0\right) = 1,$$

where $z^u(\cdot) = z_\theta^u(\cdot)$. Since

$$P\left(\max_{1 \leq j \leq N} \int_0^T |z_j^u(t)|^2\, dt < \infty,\, T > 0\right) = 1,$$

we are in the situation described in Section 4.1; thus the results of Chapter 4 are applicable here.

A control is said to be *stabilizing* if

$$P(x^u(T) \to 0 \text{ as } T \uparrow \infty) = 1.$$

The central problem of this chapter is then *to seek stabilizing admissible controls*. We make some remarks concerning the meaning and role of admissibility. First, suppose that the admissibility constraint is ignored in the search for stabilizing controls. Then the solution would be trivial: simply let $u^\#(\cdot)$ be the unique solution of the feedback law

$$P(u(t) = -F_\theta x^u(t),\, t \geq 0) = 1, \tag{1.3}$$

where F_j are matrices chosen such that $\bar{A}_j = A_j - B_j F_j$ is stable for all $j = 1, \ldots, N$. Applying 1.2.9 one ω at a time, we see that $u^\#(\cdot)$ is stabilizing.

However, for the applications motivating this chapter, this control is *useless*. This is because we interpret (1.2) to be an observation equation, that is to say that the controller has access *only* to $y^u(t), 0 \leq t \leq T$, at any given time instant T. This explains why the controls we seek must be admissible. To underscore this point, recall (4.1.9) that θ cannot be evaluated as a function of the observations up to time T for any finite T! Moreover, even though θ may sometimes be computable as a function of $y^u(t), 0 \leq t < \infty$, (consistency, Section 4.2), we cannot "wait forever" and then use (1.3): the control $u(\cdot)$ must be computed in "real time," that is, as the observations are generated.

We refer to the above problem as an *adaptive stabilization* problem. In practical terms, the admissibility constraint means that $u^\#(\cdot)$ must be the output of some (possibly nonlinear) device whose sole input is $y^u(\cdot)$. To elaborate on what such a device might be, recall that given the observations up to time T we can compute the conditional probabilities $\pi_j^u(T)$ and the state of the "jth" system $x_j^u(T)$, for each $j = 1, \ldots, N$, whenever $u(\cdot)$ is admissible.

As these are the only numerical quantities available to us, instead of (1.3) one may try

$$P(u(t) = f(x_1^u(t), \ldots, x_N^u(t), \pi_1^u(t), \ldots, \pi_N^u(t)), t \geq 0) = 1, \tag{1.4}$$

for some suitable choice of f. For example, choosing

$$f = -F \sum_{j=1}^{N} \pi_j C_j x_j$$

for some m by p matrix F yields

$$P(u(t) = -F\hat{z}^u(t), t \geq 0) = 1,$$
$$\hat{z}^u(t) = E^P(z^u(t)|\mathcal{Y}_t^u), \qquad t \geq 0. \tag{1.5}$$

Any *feedback law* of the form (1.4) actually yields a family of admissible controls, one for each set of initial conditions x^0, π^0. The goal of Section 5.3 is to establish the result that $\bar{A}_j = A_j - B_j F C_j$, $j = 1, \ldots, N$, are all stable if and only if, for each set of initial conditions, there is a unique admissible control $u^\#(\cdot)$ satisfying (1.5) that is stabilizing. Feedback laws like (1.5) may then be approximately synthesized by digital or analog circuits, and provide examples of what is meant by a "nonlinear device."

In Section 5.2 we establish the fact that given any bounded C^1 function f, the feedback (1.4) has a unique solution. The rest of the chapter is concerned with seeking stabilizing admissible controls that are optimal in a certain sense.

In Chapter 2 we were able to choose an optimal stabilizing control out of the multiplicity of all stabilizing controls by requiring that the control minimize a cost criterion. In Sections 5.4 and 5.5 we describe a similar approach; we impose a cost criterion on the class of all admissible controls *the finiteness of which implies stabilizability*; we then attempt to minimize this cost. As this cost is to be minimized only over the class of admissible controls, this problem is an example of *a partially observable optimal control problem*.

An example of such a cost is

$$E^P\left(\int_0^\infty |u(t)|^2 + |x^u(t)|^2 \, dt\right). \tag{1.6}$$

It is then an immediate consequence of 2.2.11 that any $u(\cdot)$ for which (1.6) is finite is necessarily stabilizing. The solution of the problem of minimizing (1.6) over the class of admissible controls is at present not known, and so we must choose a modified cost J^u. It turns out that a useful modification of (1.6) is to add a finite perturbation that is bounded uniformly in the control. A consequence then is that J^u is finite if and only if (1.6) is finite.

In order to motivate what we are about to add to (1.6), recall that the main obstacle toward solving these problems is the fact that $u(\cdot)$ does not have access to θ. This suggests that if $u(\cdot)$ is chosen in such a manner as best to learn what θ is, then $u(\cdot)$ may be characterizable in a simple manner. But we already have such a measure of information concerning θ: the *Shannon in-*

formation of the pair θ, $y^u(\cdot)$. Thus one approach may be to choose $u(\cdot)$ to maximize $I^u(\infty)$, the Shannon information of the pair θ, $y^u(\cdot)$. Combining this with the fact that we seek to minimize (1.6), it is natural to set

$$J^u(x^0, \pi^0) = E^P\left(\frac{1}{2}\int_0^\infty |u(t)|^2 + |x^u(t)|^2 \, dt\right) - I^u(\infty).$$

Note the presence of the minus sign in front of the information, due to the fact that we seek to minimize $J^u(x^0, \pi^0)$.

The above plausibility arguments aside, the problem of minimizing $J^u(x^0, \pi^0)$ and its analogues over the class of admissible controls is a direct generalization of the *LQ* regulator problem of Chapter 2. These *Adaptive LQ Regulator* problems will be the subject of Sections 5.4 and 5.5.

5.2. Smooth Admissible Controls

In this section we establish that given initial conditions $x_1^0, \ldots, x_N^0, \pi_1^0, \ldots,$ π_N^0 there is a unique admissible control $u^\#(\cdot)$ satisfying (1.4) whenever the feedback function is C^1 and bounded.

Let $(\Omega, \mathcal{F}, P)$ be a probability space and let $\mathcal{F}_t$, $t \geq 0$, be a nondecreasing family of sub-σ-algebras of $\mathcal{F}$. Let $\eta(\cdot)$ be an $(\Omega, \mathcal{F}_t, P)$ Brownian motion valued in $\mathbb{R}^p$ and let $\theta: \Omega \to \{1, \ldots, N\}$ be $\mathcal{F}_0$-measurable and distributed according to $\pi^0 = \{\pi_1^0, \ldots, \pi_N^0\}$.

Let $u(\cdot)$ be a control. Recall that $u(\cdot)$ is *admissible* if there are progressively measurable right continuous almost surely continuous solutions $x^u(\cdot)$, $y^u(\cdot)$ to

$$P\left(x^u(t) = x_\theta^0 + \int_0^t [A_\theta x^u(s) + B_\theta u(s)] \, ds, t \geq 0\right) = 1, \qquad (2.1)$$

$$P\left(y^u(t) = \int_0^t C_\theta x^u(s) \, ds + \eta(t), t \geq 0\right) = 1, \qquad (2.2)$$

in such a way that $u(\cdot)$ is progressively measurable relative to $\mathcal{Y}_t^u = \sigma[y^u(s), 0 \leq s \leq t], t \geq 0$.

Then for each $j = 1, \ldots, N$ there are $\mathcal{Y}_t^u$-progressively measurable right continuous almost surely continuous solutions to

$$P\left(x_j^u(t) = x_j^0 + \int_0^t [A_j x_j^u(s) + B_j u(s)] \, ds, t \geq 0\right) = 1.$$

Set $z_j^u(t) = C_j x_j^u(t), j = 1, \ldots, N, t \geq 0$. Then

$$l_j^u(T) = \exp\left(\int_0^T z_j^u(t)^* \, dy^u(t) - \frac{1}{2}\int_0^T |z_j^u(t)|^2 \, dt\right),$$

$$\pi_j^u(T) = \frac{l_j^u(T)\pi_j^0}{\sum_{k=1}^N l_k^u(T)\pi_k^0},$$

$j = 1, \dots, N$, $T \geq 0$, are *well defined*, $\mathscr{Y}_t^u$-progressively measurable, right continuous, P-almost surely continuous and

$$\pi_j^u(T) = P(\theta = j | \mathscr{Y}_T^u) \quad \text{a.s. } P \tag{2.3}$$

for all $T \geq 0$ (Theorem 4.1.2). Let

$$\hat{z}^u(T) = \sum_{k=1}^{N} \pi_k^u(T) z_k^u(T), \qquad T \geq 0;$$

then $\hat{z}^u(\cdot)$ is $\mathscr{Y}_t^u$-progressively measurable, right continuous, almost surely continuous and

$$P\left(\pi_j^u(T) = \pi_j^0 + \int_0^T \pi_j^u(t)(z_j^u(t) - \hat{z}^u(t))^*(dy^u(t) - \hat{z}^u(t) \, dt), \, T \geq 0 \right) = 1. \tag{2.4}$$

Let F be an $\mathbb{R}^m$-valued C^1 function of $x_1, \dots, x_N, \pi_1, \dots, \pi_N$. We seek admissible controls $u^\#(\cdot)$ satisfying the feedback law

$$P(u(t) = F(x_1^u(t), \dots, x_N^u(t), \pi_1^u(t), \dots, \pi_N^u(t)), \, t \geq 0) = 1. \tag{2.5}$$

Because of the possibility of explosion, $u^\#(\cdot)$ need not exist in general.

5.2.1. Theorem. *Assume that F is C^1 and bounded. Then, for each set of initial conditions, there exists a unique admissible control $u^\#(\cdot)$ that is right continuous, P-almost surely continuous, and satisfies (2.5).*

Before we give the proof, we need some preliminary facts.

5.2.2. Exercise. Let $f(\cdot)$ and $z(\cdot)$ be $\mathbb{R}$- and $\mathbb{R}^p$-valued progressively measurable processes and let $\pi(\cdot)$ be a real progressively measurable right continuous P-almost surely continuous process satisfying

$$P\left(\int_0^T |f(t)| + |z(t)|^2 \, dt < \infty, \, T > 0 \right) = 1,$$
$$P\left(\pi(t) = \int_0^t f(s)\pi(s) \, ds + \int_0^t \pi(s)z(s)^* \, d\eta(s), \, t \geq 0 \right) = 1. \tag{2.6}$$

Show that $P(\pi(t) = 0, \, t \geq 0) = 1$.

5.2.3. Exercise. Let $f(\cdot)$ and $z(\cdot)$ be $\mathbb{R}$- and $\mathbb{R}^p$-valued progressively measurable processes as above and let $\pi(\cdot)$ be a real progressively measurable right continuous P-almost surely continuous process satisfying

$$P\left(\pi(t) = \pi^0 + \int_0^t \pi(s)f(s) \, ds + \int_0^t \pi(s)z(s)^* \, d\eta(s), \, 0 \leq t < \zeta \right) = 1 \tag{2.7}$$

for some stopping time ζ. Assume that $\pi^0 \geq 0$. Show that $P(\pi(t) \geq 0, \, 0 \leq t < \zeta) = 1$.

PROOF (*of* 5.2.1). Define vector fields f, g on $\mathbb{R} \times \mathbb{R}^{n_1} \times \mathbb{R}^{n_2} \times \cdots \times \mathbb{R}^{n_N} \times \mathbb{R}^N$ by setting

$$z_j = C_j x_j, \qquad \hat{z} = \sum_{k=1}^{N} \pi_k z_k,$$

$$F = F(x_1, \ldots, x_N, \pi_1, \ldots, \pi_N),$$

$$x_j \text{ in } \mathbb{R}^{n_j}, \qquad j = 1, \ldots, N, \qquad \pi_1, \ldots, \pi_N, \lambda \text{ in } \mathbb{R},$$

$$f(\lambda, x_1, \ldots, x_N, \pi_1, \ldots, \pi_N) = \begin{pmatrix} 0 \\ A_1 x_1 + B_1 F \\ \vdots \\ A_N x_N + B_N F \\ \pi_1(z_1 - \hat{z})^*(0 - \hat{z}) \\ \vdots \\ \pi_N(z_N - \hat{z})^*(0 - \hat{z}) \end{pmatrix},$$

$$g(\lambda, x_1, \ldots, x_N, \pi_1, \ldots, \pi_N) = \begin{pmatrix} 0 \\ 0 \\ \vdots \\ 0 \\ \pi_1(z_1 - \hat{z})^* \\ \vdots \\ \pi_N(z_N - \hat{z})^* \end{pmatrix}.$$

Let φ_k be C^1 functions on $\mathbb{R}$ satisfying $\varphi_k(j) = \delta_{jk}$. Set

$$h(\lambda, x_1, \ldots, x_N, \pi_1, \ldots, \pi_N) = \sum_{k=1}^{N} \varphi_k(\lambda) C_k x_k.$$

Note then that when $\lambda = \theta$, $h = C_\theta x_\theta$.

Set

$$\xi^0 = (\theta, x_1^0, \ldots, x_N^0, \pi_1^0, \ldots, \pi_N^0).$$

Then by the existence theorem for stochastic differential equations (Section 3.2) there exists a progressively measurable right continuous P-almost surely continuous process $\xi(\cdot) = (\theta^\#(\cdot), x_1^\#(\cdot), \ldots, x_N^\#(\cdot), \pi_1^\#(\cdot), \ldots, \pi_N^\#(\cdot))$ and a stopping time ζ satisfying

$$1 = P\left(\xi(T) = \xi^0 + \int_0^T f(\xi(t)) + g(\xi(t))h(\xi(t)) \, dt \right.$$

$$\left. + \int_0^T g(\xi(t)) \, d\eta(t), 0 \le T < \zeta \right), \tag{2.8}$$

$$\lim_{t \uparrow \zeta} |\xi(t)| = +\infty \quad \text{on } \{\zeta < \infty\} \text{ a.s. } P. \tag{2.9}$$

We claim that $P(\zeta = \infty) = 1$. To this end note that by Exercise 5.2.3 we have

$$P(\pi_j^{\#}(t) \geq 0, 0 \leq t < \zeta) = 1.$$

Also $\pi(\cdot) = 1 - (\pi_1^{\#}(\cdot) + \cdots + \pi_N^{\#}(\cdot))$ satisfies an equation of the form (2.6).
By 5.2.2,

$$P\left(\sum_{k=1}^{N} \pi_k^{\#}(t) = 1, 0 \leq t < \zeta\right) = 1.$$

Since $x_j^{\#}(\cdot)$ is driven by a linear equation and F is bounded,

$$P(|x_j^{\#}(T \wedge \zeta)| < \infty, T \geq 0) = 1.$$

The form of (2.8) also implies that $P(\theta^{\#}(t) = \theta, 0 \leq t < \zeta) = 1$. Thus

$$P(|\xi(T \wedge \zeta)| < \infty, T \geq 0) = 1.$$

But this last equation coupled with (2.9) yields $P(\zeta = \infty) = 1$, which completes
the proof of the claim. Now define

$$y^{u^{*}}(t) = \int_0^t C_\theta x_\theta^{\#}(s)\, ds + \eta(t), \qquad t \geq 0.$$

Note that $u^{\#}(\cdot)$ is not defined yet, only $y^{u^{*}}(\cdot)$. Note then that, by 3.3.5, there
are $\mathcal{Y}_t^{u^{*}}$-*progressively measurable* right continuous P-almost surely continuous processes $\bar{x}_1(\cdot), \ldots, \bar{x}_N(\cdot), \bar{\pi}_1(\cdot), \ldots, \bar{\pi}_N(\cdot)$ satisfying

$$P(x_j^{\#}(t) = \bar{x}_j(t), \pi_j^{\#}(t) = \bar{\pi}_j(t), t \geq 0) = 1.$$

Set $u^{\#}(t) = F(\bar{x}_1(t), \ldots, \bar{x}_N(t), \bar{\pi}_1(t), \ldots, \bar{\pi}_N(t)), \ t \geq 0$. Then $u^{\#}(\cdot)$ is $\mathcal{Y}_t^{u^{*}}$-
progressively measurable. Set $x^{u^{*}}(t) = x_\theta^{\#}(t), \ t \geq 0$. Then the processes
$u^{\#}(\cdot), x^{u^{*}}(\cdot), y^{u^{*}}(\cdot)$ satisfy (2.1) and (2.2). We conclude that $u^{\#}(\cdot)$ is an
admissible control. In particular, $x_j^{\#}(\cdot), \pi_j^{\#}(\cdot), j = 1, \ldots, N$, and $x_j^{u^{*}}(\cdot),$
$\pi_j^{u^{*}}(\cdot), j = 1, \ldots, N$, satisfying the same set of stochastic differential equations
driven by the same control $u^{\#}(\cdot)$ in both cases. By uniqueness, it follows that

$$P(\pi_j^{\#}(t) = \pi_j^{u^{*}}(t), t \geq 0) = 1.$$

We have thus shown that $u^{\#}(\cdot)$ satisfies (2.5). Conversely, if $u(\cdot)$ satisfies (2.5),
then $(\theta, x_1^u(\cdot), \ldots, x_N^u(\cdot), \pi_1^u(\cdot), \ldots, \pi_N^u(\cdot))$ satisfies (2.8). By uniqueness

$$P(x_j^u(t) = x_j^{\#}(y), \pi_j^u(t) = \pi_j^{\#}(t), t \geq 0) = 1.$$

This implies that

$$P(u(t) = u^{\#}(t), t \geq 0) = 1. \qquad \qquad \square$$

5.2.4. Exercise. Suppose F is C^1 and bounded and let $u(\cdot)$ be an admissible
control satisfying

$$P(u(t) = F(x_1^u(t), \ldots, x_N^u(t), \pi_1^u(t), \ldots, \pi_N^u(t)), 0 \leq t < \zeta) = 1$$

for some $\mathscr{F}_t$-stopping time ζ. Show that $P(u^{\#}(t) = u(t), 0 \leq t < \zeta) = 1$.

5.3. Adaptive Stabilization

Let $(\Omega, \mathscr{F}, P)$ be a probability space and let $\mathscr{F}_t$, $t \geq 0$, be a nondecreasing family of sub-σ-algebras of $\mathscr{F}$. Let $\eta(\cdot)$ be an $(\Omega, \mathscr{F}_t, P)$ Brownian motion valued in $\mathbb{R}^p$.

For each j let n_j be a positive integer and let (A_j, B_j, C_j) be a linear system with m inputs, n_j states, and p outputs, $j = 1, \ldots, N$.

Let $\pi^0 = \{\pi_1^0, \ldots, \pi_N^0\}$ be a probability distribution on $\{1, \ldots, N\}$ and let $\theta = \theta(\pi^0): \Omega \to \{1, \ldots, N\}$ be $\mathscr{F}_0$-measurable and distributed according to π^0. Let $x^0 = \{x_1^0, \ldots, x_N^0\}$ be a choice of N initial states with x_j^0 in $\mathbb{R}^{n_j}$, $j = 1, \ldots, N$. Let $u(\cdot)$ be a control.

The control $u(\cdot)$ is *admissible at* (x^0, π^0), or (x^0, π^0)-*admissible*, if there are progressively measurable right continuous P-almost surely continuous solutions $x^u(\cdot)$, $y^u(\cdot)$ to

$$P\left(x^u(t) = x_\theta^0 + \int_0^t [A_\theta x^u(s) + B_\theta u(s)] \, ds, t \geq 0 \right) = 1, \qquad (3.1)$$

$$P\left(y^u(t) = \int_0^t C_\theta x^u(s) \, ds + \eta(t), t \geq 0 \right) = 1, \qquad (3.2)$$

in such a way that $u(\cdot)$ is progressively measurable relative to $\mathscr{Y}_t^u = \sigma[y^u(s), 0 \leq s \leq t]$, $t \geq 0$. Note that although $x^u(\cdot)$ takes values in different Euclidean spaces, the signal

$$z^u(t) = C_\theta x^u(t), \qquad t \geq 0, \qquad (3.3)$$

takes values in $\mathbb{R}^p$.

A control $u(\cdot)$ is called *stabilizing at* (x^0, π^0) if

$$P(x^u(T) \to 0 \text{ as } T\uparrow\infty) = 1. \qquad (3.4)$$

According to 5.2.1, one way to construct admissible controls is to choose a bounded C^1 function $f = f(x_1, \ldots, x_N, \pi_1, \ldots, \pi_N)$, valued in $\mathbb{R}^m$, and to let $u(\cdot)$ be the unique (x^0, π^0)-admissible control satisfying the feedback law

$$P(u(t) = f(x_1^u(t), \ldots, x_N^u(t), \pi_1^u(t), \ldots, \pi_N^u(t)), t \geq 0) = 1. \qquad (3.5)$$

Here $x_j^u(\cdot)$, $\pi_j^u(\cdot)$, $j = 1, \ldots, N$, are as in Section 5.2.

Let F be an m by p feedback matrix. We will be interested in the case of

$$f = -F \sum_{j=1}^N \pi_j C_j x_j.$$

While this f is unbounded, note that this corresponds to seeking admissible controls $u(\cdot)$ satisfying

$$P(u(t) = -F\hat{z}^u(t), t \geq 0) = 1. \qquad (3.6)$$

Here $\hat{z}^u(\cdot)$ is as in Section 5.2.

We say that the family of triples (A_j, B_j, C_j), $j = 1, \ldots, N$, is *adaptively stabilizable* if there is a fixed feedback matrix F such that for each initial condition (x^0, π^0):

(i) there is a unique (x^0, π^0)-admissible control $u^\#(\cdot)$ satisfying (3.6), and
(ii) this control $u^\#(\cdot)$ is stabilizing at (x^0, π^0).

The purpose of this section is to establish an algebraic criterion that is necessary and sufficient for adaptive stabilizability.

We say that the family of triples (A_j, B_j, C_j), $j = 1, \ldots, N$, is *uniformly stabilizable* if there is a single m by p feedback matrix F such that

$$\bar{A}_j = A_j - B_j F C_j, \qquad j = 1, \ldots, N,$$

are all stable (Section 1.2).

5.3.1. Theorem. *The family* (A_j, B_j, C_j), $j = 1, \ldots, N$, *is adaptively stabilizable by the feedback matrix F if and only if it is uniformly stabilizable by the same feedback matrix F.*

PROOF. The "only if" part is straightforward. Given $1 \leq j \leq N$, let $\pi_k^0 = \delta_{jk}$ and $\theta = \theta(\pi^0)$. Then $P(\theta = j) = 1$ and so $P(\pi_k(t) = \delta_{jk}, t \geq 0) = 1$, $k = 1, \ldots,$ N. In particular, if (3.6) held, then $P(u(t) = -FC_j x_j^u(t), t \geq 0) = 1$, which yields $P(x^u(t) = x_j^u(t), t \geq 0) = 1$. Thus $P(x_j^u(t) \to 0$ as $t \uparrow \infty) = 1$ for all x_j^0. This implies $\bar{A}_j$ is stable (1.2.9).

For the converse assume that $\bar{A}_j$ is stable for all j. We need to establish the existence (and uniqueness) of the admissible control $u^\#(\cdot)$ satisfying (3.6). Since this is somewhat technical, we assume this first, complete the proof, then come back to it.

So fix (x^0, π^0) and assume that $u(\cdot)$ is admissible and satisfies (3.6). Then

$$\begin{aligned}
\dot{x}^u &= A_\theta x^u + B_\theta u \\
&= \bar{A}_\theta x^u + B_\theta F(z^u - \hat{z}^u) \\
&= \bar{A}_\theta x^u + B_\theta F v, \qquad x^u(0) = x_\theta^0.
\end{aligned} \tag{3.7}$$

Now, by 4.3.4 and 4.3.5,

$$\tfrac{1}{2} E^P \left(\int_0^\infty |v(t)|^2 \, dt \right) = I^u(\infty) \leq -\sum_{k=1}^N \pi_k^0 \log \pi_k^0. \tag{3.8}$$

Thus, by 1.3.6, $P(x^u(t) \to 0$ as $t \uparrow \infty) = 1$. Since (x^0, π^0) is arbitrary, the result follows.

We turn now to the question of existence of $u(\cdot)$. For each $r > 0$ let $f_r = f_r(x_1, \ldots, x_N, \pi_1, \ldots, \pi_N)$ be a bounded C^1 function satisfying

$$f_r = -F \sum_{k=1}^N \pi_k C_k x_k \qquad \text{on} \qquad \sum_{k=1}^N |x_k| \leq r.$$

Let $u^r(\cdot)$ be the admissible control satisfying (3.5) with $f = f_r$. Let the quantities associated to $u^r(\cdot)$ be denoted $x^r(\cdot)$, $x^r_j(\cdot)$, $\pi^r_j(\cdot)$. Let

$$\tau^r = \inf\{t \geq 0 \,||\, x^r_1(t)| + \cdots + |x^r_N(t)| \geq r\}.$$

Then, by 5.2.4,

$$u^r(t) = u^{r+1}(t) \qquad \text{on} \quad \{t < \tau^r \wedge \tau^{r+1}\} \quad \text{a.s. } P$$

which implies

$$x^r_j(t) = x^{r+1}_j(t) \qquad \text{on} \quad t < \tau^r \wedge \tau^{r+1} \quad \text{a.s. } P.$$

Hence $P(\tau^r \leq \tau^{r+1}) = 1$. Now note that $P(u^r(t) = -F\hat{z}^r(t), 0 \leq t < \tau^r) = 1$. Thus $x^r(\cdot)$ satisfies an equation analogous to (3.7) up to time τ^r. This fact with the estimate analogous to (3.8) implies that *a priori*

$$|x^r(T \wedge \tau^r)|^2 \leq c_T \times \left(1 + \int_0^{T \wedge \tau^r} |z^r(t) - \hat{z}^r(t)|^2 \, dt\right)$$

and

$$E^P(|x^r(T \wedge \tau^r)|^2) \leq c_T \times \left(1 - \sum_{k=1}^{N} \pi^0_k \log \pi^0_k\right) \tag{3.9}$$

for all $T > 0$, where c_T is a constant depending on T.

Let $\tau = \lim_{r \to \infty} \tau^r \leq +\infty$. Then $\tau \leq T$ implies that for all r we have $|x^r(\tau^r)| = r$. But, by (3.9), the probability of this last happening is zero. Thus $P(\tau = \infty) = 1$. Now let $u(\cdot)$, $x_j(\cdot)$, $\pi_j(\cdot)$ be $\mathscr{F}_t$-progressively measurable right continuous processes that equal $u^r(\cdot)$, $x^r_j(\cdot)$, $\pi^r_j(\cdot)$ on $\{t < \tau^r\}$ for all $r = 1, 2, 3, \ldots$. By 3.1.15, these processes are well defined. Then $\xi(\cdot) = (\theta, x_1(\cdot), \ldots, x_N(\cdot), \pi_1(\cdot), \ldots, \pi_N(\cdot))$ satisfies (2.8) with $\zeta = \infty$. The technique used in the second half of the proof of 5.2.1 implies that there is a control $\bar{u}(\cdot)$ that is *admissible* and equalling $u(\cdot)$ a.s. P. We have thus constructed an admissible control $\bar{u}(\cdot)$ that satisfies (3.6). Uniqueness follows now as in the proof of uniqueness in 5.2.1. This completes the proof. $\qquad \square$

The reader should note that the case of the matrices (A_j, B_j, C_j) being independent of j still contains something of interest: this corresponds to the case of a *known* linear system with an *unknown* (initial) state. The next exercise gives another important special case.

5.3.2. Exercise. Let (A_j, B_j, C_j) be a minimal triple for each $j = 1, \ldots, N$. Set $F_j = B^*_j K_j$, $K_j = K(A_j, B_j, C_j)$, $j = 1, \ldots, N$. Show that the family (A_j, B_j, F_j), $j = 1, \ldots, N$, is uniformly stabilizable with $F = I$.

It is the corresponding family of stabilizing feedback laws that will be of interest in the next section.

We note that when the matrices C_j, $j = 1, \ldots, N$, are equal to the identity $(n = p)$, whether or not the family of pairs (A_j, B_j), $j = 1, \ldots, N$, is uniformly stabilizable is a "completely observable" property whereas adaptive stabilizability is, even in this case, still a "partially observable" property.

5.3.3. Exercise. Fix (x^0, π^0) and let $u(\cdot)$ be an (x^0, π^0)-admissible control satisfying (3.6). Show that the corresponding cost (1.6) is *finite*. (*Hint*: Use 1.3.8 with $r = 2$.)

5.4. Optimal Control

The purpose of this section is to show that, in certain cases, one can exhibit not only admissible controls that are stabilizing but also ones that are stabilizing and *optimal* in the sense that they minimize a cost criterion similar to (1.6).

To conform with the notation previously established in Chapter 2, the notation here and in the next section differs slightly from that of Sections 5.1, 5.2, and 5.3. The main result of this section, Theorem 5.4.5, can be viewed as a direct generalization of the case $N = 1$ dealt with in Chapter 2.

Let $(\Omega, \mathcal{F}, P)$ be a probability space and let $\mathcal{F}_t$, $t \geq 0$, be a nondecreasing family of sub-σ-algebras of $\mathcal{F}$. Let $\eta(\cdot)$ be an $(\Omega, \mathcal{F}_t, P)$ Brownian motion valued in $\mathbb{R}^m$.

For each $j = 1, \ldots, N$ let n_j be a positive integer and let (A_j, B_j, C_j) be a linear system with m inputs, n_j states, and p outputs. For each $j = 1, \ldots, N$ let F_j be an m by n_j matrix and let x_j^0 be an initial state in $\mathbb{R}^{n_j}$; let x^0 denote $\{x_1^0, \ldots, x_N^0\}$. Let $\pi^0 = \{\pi_1^0, \ldots, \pi_N^0\}$ be a distribution on $\{1, \ldots, N\}$ and let $\theta = \theta(\pi^0): \Omega \to \{1, \ldots, N\}$ be $\mathcal{F}_0$-measurable and distributed according to π^0.

A control $u(\cdot)$ is (x^0, π^0)-*admissible* provided there are progressively measurable right continuous *P*-almost surely continuous solutions $x^u(\cdot)$, $y^u(\cdot)$ to

$$P\left(x^u(t) = x_\theta^0 + \int_0^t [A_\theta x^u(s) + B_\theta u(s)]\, ds, t \geq 0\right) = 1, \qquad (4.1)$$

$$P\left(y^u(t) = \int_0^t F_\theta x^u(s)\, ds + \eta(t), t \geq 0\right) = 1 \qquad (4.2)$$

in such a way that $u(\cdot)$ is progressively measurable relative to $\mathcal{Y}_T^u = \sigma[y^u(t), 0 \leq t \leq T]$, $T \geq 0$. Note that the matrices C_j do not appear in (4.2); they appear instead in the cost J^u defined below.

With $x_j^u(\cdot)$ as in Section 5.2, we set $z_j^u(\cdot) = F_j x_j^u(\cdot)$ and $z^u(\cdot) = z_\theta^u(\cdot)$, $j = 1, \ldots, N$. Then the processes $z_j^u(\cdot)$ are progressively measurable relative to $\mathcal{Y}_t^u$, $t \geq 0$, for each $j = 1, \ldots, N$, whenever $u(\cdot)$ is admissible. Let $\pi_j^u(\cdot)$ and $\hat{z}^u(\cdot)$ be defined now as in Section 5.2. Set

$$v^u(t) = y^u(t) - \int_0^t \hat{z}^u(s)\, ds, \qquad t \geq 0;$$

then, according to 4.1.6, $v^u(\cdot)$ is an $(\Omega, \mathcal{Y}_t^u, P)$ Brownian motion, whenever $u(\cdot)$ is admissible. Moreover, in this case one has ((1.12) of Chapter 4)

$$P\left(\pi_j^u(t) = \pi_j^0 + \int_0^t \pi_j^u(s)(z_j^u(s) - \hat{z}^u(s))^*\, dv^u(s), t \geq 0\right) = 1. \qquad (4.3)$$

Let $I^u(\infty)$ denote the Shannon information (Section 4.3) of the pair θ, $y^u(t)$, $0 \leq t < \infty$, and set

$$J^u(x^0, \pi^0) = \tfrac{1}{2}E^P\left(\int_0^\infty |u(t)|^2 + |C_\theta x^u(t)|^2 \, dt\right) - I^u(\infty), \qquad \theta = \theta(\pi^0).$$

$$(4.4)$$

This is the *cost* corresponding to the control $u(\cdot)$ and the initial conditions (x^0, π^0). By 4.3.4,

$$+\infty \geq J^u(x^0, \pi^0) \geq \sum_{k=1}^N \pi_k^0 \log \pi_k^0 > -\infty.$$

We seek to minimize $J^u(x^0, \pi^0)$ over the class of controls admissible at (x^0, π^0). Set

$$S(x^0, \pi^0) = \inf\{J^u(x^0, \pi^0)|u(\cdot) \text{ admissible at } (x^0, \pi^0)\}. \qquad (4.5)$$

Recall that for a minimal triple (A, B, C), $K = K(A, B, C)$ denotes the positive definite solution of the algebraic Riccati equation (Section 2.3).

5.4.1. Lemma. *Assume that the triple (A_j, B_j, C_j) is minimal for each $j = 1, \ldots,$ N. Then $(K_j = K(A_j, B_j, C_j))$*

$$S(x^0, \pi^0) \geq \sum_{j=1}^N \pi_j^0(\tfrac{1}{2}x_j^0*K_j x_j^0 + \log \pi_j^0).$$

PROOF. Follows from (4.5), 2.3.8, and 4.3.4. ☐

A control $u^\#(\cdot)$ is *optimal* at (x^0, π^0) if $S(x^0, \pi^0) = J^{u^\#}(x^0, \pi^0)$. The problem in this section is to seek a control $u^\#(\cdot)$ that is admissible and optimal at (x^0, π^0). Such a control is then necessarily stabilizing, provided $S(x^0, \pi^0) < \infty$, according to the following lemma.

5.4.2. Lemma. *Assume that (A_j, B_j, C_j) is minimal for each $j = 1, \ldots, N$, and suppose that $J^u(x^0, \pi^0) < \infty$. Then $u(\cdot)$ is stabilizing at (x^0, π^0).*

PROOF. By 2.3.8, $(K_\theta = K(A_\theta, B_\theta, C_\theta))$

$$P\left(x^u(T)*K_\theta x(T) \leq \int_T^\infty |u(t)|^2 + |C_\theta x^u(t)|^2 \, dt, \ T \geq 0\right) = 1.$$

The finiteness of $J^u(x^0, \pi^0)$ then implies that the integral goes to zero as $T \uparrow \infty$. The result follows. ☐

In this section we shall succeed in finding an expliciting expression for the optimal control only under very stringent hypotheses. We shall assume:

(A1) For each $j = 1, \ldots, N$ the triple (A_j, B_j, C_j) is minimal.
(A2) For each $j = 1, \ldots, N$, $F_j = B_j^* K_j$.

Two remarks are in order. First, (A2) will be generalized slightly in the next section. Second, since one can choose C_j arbitrarily (subject to (A1)), there is some leeway in what values the matrices F_j may take. Assumptions (A1) and (A2) imply, according to 5.3.2, that the family of triples $(A_j, B_j, F_j), j = 1, \ldots, N$, is a uniformly stabilizable family with $F = I$. In Section 5.3 we have seen how to construct an admissible stabilizing control in this case. The main result of this section shows that the control constructed in Section 5.3 is actually optimal relative to the cost (4.4). The key computation here is the "completing-the-square" argument in 5.4.4.

For any random time τ set

$$I^u(\tau) = \tfrac{1}{2}E^P\left(\int_0^\tau |z^u(t) - \hat{z}^u(t)|^2\, dt\right) \tag{4.6}$$

and

$$J_\tau^u(x^0, \pi^0) = \tfrac{1}{2}E^P\left(\int_0^\tau |u(t)|^2 + |C_\theta x^u(t)|^2\, dt\right) - I^u(\tau).$$

Then, by 4.3.5, $I^u(\tau) \to I^u(\infty)$ as $\tau\uparrow\infty$, and hence $J_\tau^u(x^0, \pi^0) \to J^u(x^0, \pi^0)$ as $\tau\uparrow\infty$ for any admissible control $u(\cdot)$. If $\tau = T$, then $I^u(\tau)$ is the Shannon information of pair $\theta, y^u(t), 0 \le t \le T$.

5.4.3. Lemma. *For any control* $u(\cdot)$,

$$2J_\tau^u(x^0, \pi^0) + E^P(x^u(\tau)^*K_\theta x^u(\tau) - x_\theta^{0*}K_\theta x_\theta^0)$$
$$= E^P\left(\int_0^\tau |u(t) + z^u(t)|^2\, dt\right) - 2I^u(\tau). \tag{4.7}$$

PROOF. Follows immediately from 2.3.4.   □

As in Section 5.3, for each $r > 0$ let

$$\tau_r = \inf\{t \ge 0 \,|\, |z_1^u(t)| + \cdots + |z_N^u(t)| \ge r\}.$$

Then, as in the proof of 4.1.6, for Φ integrable and $\mathcal{Y}_t^u$-measurable and $T > 0$,

$$E^P(\Phi(z^u(t) - \hat{z}^u(t)), t < T \wedge \tau_r) = 0. \tag{4.8}$$

In particular, (4.8) holds with $\Phi = u(t)^* + \hat{z}^u(t)^*$ whenever $u(\cdot)$ is admissible and $E^P(|u(t)|) < \infty$.

5.4.4. Proposition. *Let* $u(\cdot)$ *be admissible at* (x^0, π^0) *and satisfy* $J^u(x^0, \pi^0) < \infty$. *Then*

$$J^u(x^0, \pi^0) = \tfrac{1}{2}E^P\left(x_\theta^{0*}K_\theta x_\theta^0 + \int_0^\infty |u(t) + \hat{z}^u(t)|^2\, dt\right). \tag{4.9}$$

PROOF. We use (4.6), (4.7), and (4.8). Since J^u is finite we have $E^P(|u(t)|) < \infty$ for (Lebesgue) almost all $t \ge 0$. Thus for almost all $t \ge 0$ (4.8) holds with

$\Phi = u(t)^* + \hat{z}^u(t)^*$. This implies $(\tau = T \wedge \tau_r)$

$2J_\tau^u(x^0, \pi^0) + E^P(x^u(\tau)^* K_\theta x^u(\tau) - x_\theta^{0*} K_\theta x_\theta^0)$

$$= E^P\left(\int_0^\tau |u + z^u|^2 \, dt\right) - 2I^u(\tau) \qquad \text{(by (4.7))}$$

$$= E^P\left(\int_0^\tau [|u|^2 + 2u^*z^u + |z^u|^2 - |z^u|^2 + 2z^{u*}\hat{z}^u - |\hat{z}^u|^2] \, dt\right) \qquad \text{(by (4.6))}$$

$$= E^P\left(\int_0^\tau [|u|^2 + 2u^*\hat{z}^u + |\hat{z}^u|^2 + 2\Phi(z^u - \hat{z}^u)] \, dt\right)$$

$$= E^P\left(\int_0^\tau [|u|^2 + 2u^*\hat{z}^u + |\hat{z}^u|^2] \, dt\right) \qquad \text{(by (4.8))}$$

$$= E^P\left(\int_0^\tau |u + \hat{z}^u|^2 \, dt\right). \qquad (4.10)$$

Now 5.4.2 also shows that $x^u(\tau)^* K_\theta x^u(\tau)$, $\tau \geq 0$, are all bounded by the same integrable random variable and so, by letting $r \uparrow \infty$, we see that (4.10) holds with τ replaced by T. Now if we let $T \uparrow \infty$, the same argument applies and so (4.10) holds with $T = \infty$. But 5.4.2 says that $u(\cdot)$ is stabilizing and so the result follows. $\qquad \square$

Note how (A2) is used crucially in the derivation of (4.9).

5.4.5. Theorem. *Assume* (A1) *and* (A2) *and fix* (x^0, π^0). *Then there is a unique* (x^0, π^0)-*admissible control* $u^\#(\cdot)$ *satisfying the feedback law*

$$P(u(t) = -\hat{z}^u(t), t \geq 0) = 1.$$

Moreover, $u^\#(\cdot)$ *is the unique* (x^0, π^0)-*admissible control that is optimal at* (x^0, π^0), *and*

$$S(x^0, \pi^0) = J^{u^\#}(x^0, \pi^0) = E^P(\tfrac{1}{2}x_\theta^{0*} K_\theta x_\theta^0).$$

PROOF. Since (A1) and (A2) imply that the family of triples (A_j, B_j, F_j), $j = 1, \ldots, N$, is uniformly stabilizable with $F = I$, the existence and uniqueness of an admissible control $u^\#(\cdot)$ satisfying the stated feedback law follows from 5.3.1. Moreover, Exercise 5.3.3 implies that the corresponding cost (1.6) is finite. This implies that $J^u(x^0, \pi^0)$ is finite. Thus (4.9) holds and so $u^\#(\cdot)$ is clearly optimal. Since any other optimal control necessarily satisfies this feedback law, $u^\#(\cdot)$ is the unique optimal law. Moreover, (4.9) yields the expression for the optimal cost. $\qquad \square$

In the next section we rederive this theorem in a different manner, emphasizing the connection with a nonlinear differential equation called the Bellman equation. As a dividend, we are able to prove that this Bellman equation has a unique solution related to the function S exhibited above.

5.5. Bellman Equation

We follow the notation established in the previous section: for each $j = 1, \ldots,$ N we are given a linear system (A_j, B_j, C_j) with m inputs, n_j states, and p outputs, a q by n_j matrix F_j, and a positive real number q_j. In this section we assume that:

(A1) (A_j, B_j, C_j) is minimal for all $j = 1, \ldots, N$,
(A2) $F_j^* F_k = (1/q_j q_k) K_j B_j B_k^* K_k$ with $K_j = K(A_j, B_j/\sqrt{q_j}, C_j)$ for all $j, k = 1,$ $\ldots, N$.

Then (A2) appearing in the previous section is a special case of (A2) here. For each distribution $\pi = \{\pi_1, \ldots, \pi_N\}$ let

$$H(\pi) = -\sum_{k=1}^{N} q_k \pi_k \log \pi_k + Q \log Q, \qquad Q = \sum_{k=1}^{N} q_k \pi_k.$$

Set

$$J^u(x^0, \pi^0) = E^P\left(\frac{1}{2}\int_0^\infty q_\theta |u(t)|^2 + |C_\theta x^u(t)|^2 \, dt + H(\pi^u(\infty))\right), \qquad \theta = \theta(\pi^0).$$

Recall that $x_j^u(\cdot), \pi_j^u(\cdot), j = 1, \ldots, N$, satisfy

$$P\left(x_j^u(t) = x_j^0 + \int_0^t [A_j x_j^u(s) + B_j u(s)] \, ds, t \geq 0\right) = 1,$$

$$P\left(\pi_j^u(t) = \pi_j^0 + \int_0^t \pi_j^u(s)(z_j^u(s) - \hat{z}^u(s))^* \, dv^u(s), t \geq 0\right) = 1,$$

$j = 1, \ldots, N$, where $z_j^u(\cdot) = F_j x_j^u(\cdot)$, $\hat{z}^u(\cdot) = \sum_{k=1}^{N} \pi_k^u(\cdot) z_k^u(\cdot)$, and $v^u(\cdot)$ is the innovations process, an $(\Omega, \mathcal{Y}_t, P)$ Brownian motion, for all admissible controls $u(\cdot)$.

Let $\mathcal{P}$ denote the set of all distributions π on $\{1, \ldots, N\}$ satisfying $\pi_j > 0$ for all j. Let $\mathcal{D} = \mathbb{R}^{n_1} \times \cdots \times \mathbb{R}^{n_N} \times \mathcal{P}$, and let $\bar{\mathcal{D}}$ denote the closure of $\mathcal{D}$ in the ambient Euclidean space. A function $f: \mathcal{D} \to \mathbb{R}$ is said to be in $C^k(\mathcal{D})$ if the function

$$\bar{f}(x_1, \ldots, x_N, \pi_1, \ldots, \pi_{N-1}) = f\left(x_1, \ldots, x_N, \pi_1, \ldots, \pi_{N-1}, 1 - \sum_{k=1}^{N-1} \pi_k\right)$$

is C^k in the usual sense for all x_j in $\mathbb{R}^{n_j}$ and $0 < \sum_{k=1}^{N-1} \pi_k < 1$. For f in $C^2(\mathcal{D})$ set

$$G^0(f) = \frac{1}{2}\sum_{j,k=1}^{N-1} \frac{\partial^2 \bar{f}}{\partial \pi_j \partial \pi_k} \pi_j \pi_k (z_j - \hat{z})^*(z_k - \hat{z}) + \sum_{j=1}^{N} \frac{\partial f}{\partial x_j} A_j x_j,$$

where $z_j = F_j x_j$ and $\hat{z} = \sum_{j=1}^{N} \pi_j z_j$. For u in $\mathbb{R}^m$ set

$$G^u(f) = \sum_{j=1}^{N} \frac{\partial f}{\partial x_j} B_j u + G^0(f).$$

Then f in $C^2(\mathcal{D})$ implies $G''(f)$ is in $C^0(\mathcal{D})$. A function f is said to be in $C^0(\bar{\mathcal{D}})$ if f extends to a continuous function on $\bar{\mathcal{D}}$. The function f is in $C^1(\bar{\mathcal{D}})$ if $\partial f/\partial x_j$, $j = 1, \ldots, N$, $\partial f/\partial \pi_k$, $k = 1, \ldots, N - 1$, extend to continuous functions on $\{(x_1, \ldots, x_N, \pi_1, \ldots, \pi_{N-1}) | x_j \text{ in } \mathbb{R}^{n_j}, 0 \le \pi_j \le 1, \text{ and } 0 \le \pi_1 + \cdots + \pi_{N-1} \le 1\}$. In this case it can be shown that there is a C^1 function F on $\mathbb{R}^{n_1} \times \cdots \times \mathbb{R}^{n_N} \times \mathbb{R}^N$ such that F is bounded on $\sum_{j=1}^N |x_j| \le r$ for all r and $f = F$ on $\mathcal{D}$. For example, the function

$$S(x, \pi) = \sum_{j=1}^N \pi_j x_j^* K_j x_j + H(\pi) \tag{5.1}$$

is in $C^0(\bar{\mathcal{D}}) \cap C^2(\mathcal{D})$ and $\partial S/\partial x_j, j = 1, \ldots, N$, are in $C^1(\bar{\mathcal{D}})$.

In this section we derive the following results.

5.5.1. Theorem. *There is a unique admissible control $u^{\#}(\cdot)$ satisfying*

$$P\left(u(t) = -\frac{1}{Q(\pi''(t))} \sum_{j=1}^N \pi_j''(t) B_j^* K_j x_j''(t), t \ge 0\right) = 1.$$

This control is optimal, i.e., $J^{u^{\#}}(x^0, \pi^0) < J^u(x^0, \pi^0)$ for all other admissible $u(\cdot)$, and $J^{u^{\#}}(x^0, \pi^0) = S(x^0, \pi^0)$, where S is given by (5.1).

5.5.2. Theorem. *The function S given by (5.1) is the unique function in $C^2(\mathcal{D})$ with $\partial S/\partial x_j, j = 1, \ldots, N$, all in $C^1(\bar{\mathcal{D}})$ and satisfying the Bellman equation*

$$G^0(S)(x, \pi) = \frac{1}{2Q(\pi)}\left|\sum_{j=1}^N \frac{\partial S}{\partial x_j}(x, \pi) B_j\right|^2 - \frac{1}{2}\sum_{j=1}^N \pi_j |C_j x_j|^2 \tag{5.2}$$

together with the growth condition

$$H(\pi) \le S(x, \pi) \le \text{constant} \times \sum_{j=1}^N \pi_j |x_j|^2 + H(\pi) \tag{5.3}$$

for all (x, π) in $\mathcal{D}$.

It can be shown, by direct calculation, that the function S given by (5.1) does indeed satisfy (5.2) upon appealing to the definition of K_j. We prove the above theorems by establishing three lemmas. In fact, the special case $N = 1$ of (5.2) yields the algebraic Riccati equation upon the substitution $S = x^* K x/2$.

5.5.3. Lemma. *Let S be any function in $C^2(\mathcal{D})$ with $\partial S/\partial x_j, j = 1, \ldots, N$, all in $C^1(\bar{\mathcal{D}})$ and satisfying (5.2) and (5.3). Then for (x^0, π^0) in $\mathcal{D}$ and $u(\cdot)$ (x^0, π^0)-admissible with $J^u(x^0, \pi^0) < \infty$,*

$$J^u(x^0, \pi^0) = S(x^0, \pi^0) + \frac{1}{2}E^P\left(\int_0^\infty Q\left|u^* + \frac{1}{Q}\sum_{j=1}^N \frac{\partial S}{\partial x_j} B_j\right|^2 dt\right),$$

where $Q = Q(\pi''(t))$ and u^* denotes the transpose of u.

PROOF. For $f: \mathscr{D} \to \mathbb{R}$ set $f(\cdot) = f(x_1^u(\cdot), \ldots, x_N^u(\cdot), \pi_1^u(\cdot), \ldots, \pi_N^u(\cdot))$. Apply the Ito rule (Section 3.2) to $S(\cdot)$ to yield

$$S(t) - S(x^0, \pi^0) - \int_0^t G^{u(r)}(S)(r) \, dr, \qquad t \geq 0, \tag{5.4}$$

is an $(\Omega, \mathscr{Y}_t, P)$ stochastic integral. Let τ_n be the smallest of the first times T that T, $\int_0^T |u(t)|^2 \, dt$, $|x_j^u(T)|$, or $\pi_j^u(T)^{-1}$ exceeds $n \geq 1$. Then the integrand appearing in the stochastic integral (5.4) is bounded up to time τ_n and thus

$$0 = S(x^0, \pi^0) + E^P\left(-S(\tau_n) + \int_0^{\tau_n} G^{u(t)}(S)(t) \, dt\right) \tag{5.5}$$

for all $n \geq 1$. Since (x^0, π^0) is in $\mathscr{D}$, we have $P(\tau_n \uparrow \infty \text{ as } n \uparrow \infty) = 1$. Set $F = S - H$. Then

$$0 \leq E^P(F(\tau_n)) \leq C \times E^P\left(\sum_{j=1}^N \pi_j^u(\tau_n)|x_j^u(\tau_n)|^2\right)$$

$$= C \times E^P(|x_\theta^u(\tau_n)|^2)$$

$$\leq C' \times E^P\left(\int_{\tau_n}^\infty q_\theta |u(t)|^2 + |C_\theta x^u(t)|^2 \, dt\right).$$

Thus $\lim_{n \uparrow \infty} E^P(F(\tau_n)) = 0$. Now appealing to (5.2), (5.5), and completing the square, we have

$$E^P\left(\tfrac{1}{2} \int_0^{\tau_n} q_\theta |u(t)|^2 + |C_\theta x^u(t)|^2 \, dt + H(\pi^u(\tau_n))\right)$$

$$= E^P\left(\tfrac{1}{2} \int_0^{\tau_n} Q(\pi^u(t))|u(t)|^2 + \sum_{j=1}^N \pi_j^u(t)|C_j x_j^u(t)|^2 \, dt + H(\pi^u(\tau_n))\right)$$

$$= S(x^0, \pi^0) + E^P\left(-F(\tau_n) + \tfrac{1}{2} \int_0^{\tau_n} Q\left|u^* + \frac{1}{Q} \sum \frac{\partial S}{\partial x_j} B_j\right|^2 \, dt\right) \tag{5.6}$$

for all $n \geq 1$. The result follows upon letting $n \uparrow \infty$. $\square$

5.5.4. Lemma. *Let S be as in 5.5.3. Then for each (x^0, π^0) in $\mathscr{D}$ there exists a unique admissible control $u^\#(\cdot)$ satisfying*

$$P\left(u(t)^* = -\frac{1}{Q} \sum_{j=1}^N \frac{\partial S}{\partial x_j}(x^u(t), \pi^u(t))B_j, t \geq 0\right) = 1, \qquad Q = Q(\pi^u(t)). \tag{5.7}$$

Moreover, $J^{u^\#}(x^0, \pi^0) = S(x^0, \pi^0)$ is finite.

PROOF. For each $n \geq 1$, choose f_n such that

$$f_n = -\frac{1}{Q} \sum_{j=1}^N \frac{\partial S}{\partial x_j}(x, \pi)B_j \qquad \text{on} \qquad \sum_{j=1}^N |x_j| \leq n$$

and f_n is C^1 and bounded. We follow the proof of 5.3.1. Let $u^n(\cdot)$ be the admissible control satisfying (3.5) with $f = f_n$. Let

$$\tau^n = \inf\left\{t \geq 0 \;\middle|\; \sum_{j=1}^N |x_j^n(t)| \geq n\right\}.$$

Then, as in 5.3.1, we have $P(\tau^n \leq \tau^{n+1}) = 1$ and $u^n(\cdot)$ satisfies

$$P\left(u(t)^* = -\frac{1}{Q} \sum_{j=1}^N \frac{\partial S}{\partial x_j}(x^n(t), \pi^n(t))B_j,\ 0 \leq t < \tau^n\right) = 1.$$

Now an easy modification of the argument in 5.5.3 shows that (5.6) also holds for the present choice of stopping times τ^n; thus one has the *a priori* estimate

$$E^P\left(\frac{1}{2}\int_0^{T \wedge \tau^n} |u^n(t)|^2\, dt\right) \leq \frac{S(x^0, \pi^0)}{\min(q_1, \ldots, q_N)} < \infty,$$

since $F \geq 0$. Now this estimate implies, by the same procedure as in 5.3.1, that $\tau = \lim_{n \uparrow \infty} \tau^n = \infty$ P-almost surely and the existence of a unique admissible control $u^\#(\cdot)$ satisfying (5.7). Appealing to (5.6) once more yields the fact that $J^{u^\#}(x^0, \pi^0) \leq S(x^0, \pi^0)$; the result follows.   □

5.5.5. Lemma. *With S as in 5.5.3,*

$$S(x^0, \pi^0) = \inf\{J^u(x^0, \pi^0)| all\ u(\cdot)\ admissible\ at\ (x^0, \pi^0)\}$$

for all (x^0, π^0) in $\mathscr{D}$.

PROOF. By 5.5.3, $J^u(x^0, \pi^0) \geq S(x^0, \pi^0)$ for all admissible $u(\cdot)$. By 5.5.4, $J^{u^\#}(x^0, \pi^0) = S(x^0, \pi^0)$. The result follows.   □

The above three lemmas combined yield the proofs of Theorems 5.5.1 and 5.5.2.   □

Note that, in particular, the control whose existence is asserted in 5.5.1 is of course stabilizing,

$$P(x^{u^\#}(t) \to 0 \text{ as } t \uparrow \infty) = 1.$$

5.5.6. Exercise. This provides an example of what may happen in case (A2) does not hold. Take $N = 2$, $m = n_1 = n_2 = p = q = 1$, $a_1 = a_2 = 0$, and $b_1 = -b_2 = c_1 = c_2 = f_1 = f_2 = q_1 = q_2 = 1$. Show that if $x_1^0 = x_2^0 = x^0$ and $\pi_1^0 = \pi_2^0$, the control $u^\#(\cdot) = 0$ identically satisfies the feedback law appearing in 5.5.1. In particular, $u^\#(\cdot)$ is not stabilizing if $x^0 \neq 0$ and so $J^{u^\#}(x^0, x^0) = +\infty$. Show that, in spite of this, the function $S = (\pi_1 x_1^2 - \pi_2 x_2^2)/2 + H(\pi_1, \pi_2)$ satisfies the Bellman equation and is in $C^2(\mathscr{D})$ with $\partial S/\partial x_1$, $\partial S/\partial x_2$ in $C^1(\bar{\mathscr{D}})$. The only condition that fails here is the *positivity condition* (5.3).

5.6. Notes and References

The term "Adaptive Control" covers an extensive field in which a wide variety of techniques are brought to bear on the intuitively appealing problem of stabilizing an "uncertain" system. The approach taken in this chapter is only one of many, and concerns itself with a specific form of the general problem, the Adaptive *LQ* Regulator. This type of problem was first addressed in [5.2] in the early 1960s; since then it has not attracted much attention, due perhaps to the fact that the relevant filtering results were unavailable till the 1970s. Reference [4.3] considers a finite-time horizon version of this problem. A good survey of the field of "Adaptive Control" is [5.5]. In [5.1] a Gaussian non-adaptive "*LQG*" version is described. In [5.4] the "*LQG*" time-average version of the results of this chapter is described.

The concept of uniform stabilizability is known in the literature as "simultaneous stabilizability by static feedback." A more general form of feedback, "dynamic compensation," is the subject of [5.3]. A family of triples (A_j, B_j, C_j) is "simultaneously stabilizable by dynamic compensation" if there is a transfer function $F(s)$ such that with $G_j(s) = C_j(sI - A_j)^{-1}B_j$, the systems $\bar{G}_j(s) = G_j(s)(I + F(s)G_j(s))^{-1}$ are stable for all $j = 1, \ldots, N$. It is known that this is possible in many cases [5.3]: the corresponding generalization of Theorem 5.3.1 is open. The results of Sections 5.4 and 5.5 appear (with $q_1 = q_2 = \cdots = q_N = 1$) in [5.4]. The example in Exercise 5.5.6 is due to Y. Bar-Shalom (private communication).

[5.1]   M. H. A. Davis, *Linear Estimation and Stochastic Control*, Chapman and Hall, London, 1977.
[5.2]   A. A. Fel'dbaum, *Optimal Control Systems*, Academic Press, New York, 1965.
[5.3]   B. K. Ghosh and C. I. Byrnes, "Simultaneous Stabilization and Simultaneous Pole Placement by Nonswitching Dynamic Compensation," *IEEE Trans. Automat. Control*, **28** (1983), 735–741.
[5.4]   O. Hijab, "The Adaptive *LQG* Problem, II," *Stochastics*, **16** (1986), 25–49.
[5.5]   P. R. Kumar, "A Survey of Some Results in Stochastic Adaptive Control," *SIAM J. Control Optim.*, **23** (1985), 329–380.

APPENDIX
Solutions to Exercises

Chapter 1

1.2.1. Note first that (2.3) implies the validity of the statements when f and g are polynomials. Fix a contour C containing the eigenvalues of A and let k_A be the maximum of $|(sI - A)^{-1}|$ (length(C))/2π as s varies over C. Then for any entire h one has $|h(A)| \leq k_A \max_C |h|$. Let $p_n(s)$, $q_n(s)$ be polynomials such that p_n and q_n converge uniformly on C to f and g, respectively, as $n \uparrow \infty$. Then $p_n q_n$ and $p_n + q_n$ converge uniformly on C to fg and $f + g$, respectively. Thus $(fg)(A) = \lim_{n \uparrow \infty}(p_n q_n)(A) = \lim_{n \uparrow \infty} p_n(A)q_n(A) = \lim_{n \uparrow \infty} p_n(A)\lim_{n \uparrow \infty} q_n(A) = f(A)g(A)$. The other statement follows similarly.

1.2.2. Let $v \neq 0$ satisfy $Av = \lambda v$. Then $m(\lambda)v = m(A)v = 0$; thus $m(\lambda) = 0$. If $p(s)(sI - A)^{-1}$ has polynomial entries then by (2.1) we have $p(A) = 0$. Conversely, observe first that

$$\frac{p(s) - p(t)}{s - t} = \sum_{k=0}^{n} j_k(t)s^k$$

for some polynomials $j_k(t)$, $k = 0, \ldots, n$. Plugging in $t = A$ and appealing to 1.2.1 yields

$$(p(s)I - p(A))(sI - A)^{-1} = \sum_{k=0}^{n} j_k(A)s^k.$$

This shows that $p(A) = 0$ implies that $p(s)(sI - A)^{-1}$ has polynomial entries.

1.2.3. Choose a basis for $\mathbb{C}^n$ by first choosing one for V and then extending it to all of $\mathbb{C}^n$. In this basis the matrix of A looks like

$$\bar{A} = \begin{pmatrix} B & 0 \\ C & D \end{pmatrix},$$

where B is the matrix corresponding to $A|_V$. Since $\det(sI - \bar{A}) = \det(sI - B)\det(sI - D)$, the result follows.

1.2.4. Use 1.2.7: first,

$$(sI - A)^{-1} = \frac{1}{q(s)} \begin{pmatrix} s(s^2 + 4\omega^2) & s^2 & 0 & 2\omega\sigma s \\ 3\omega^2 s^2 & s^3 & 0 & 2\omega\sigma s^2 \\ -\dfrac{6\omega^3}{\sigma} & -\dfrac{2\omega s}{\sigma} & s(s^2 + \omega^2) & s^2 - 3\omega^2 \\ -\dfrac{6\omega^3 s}{\sigma} & -\dfrac{2\omega s^2}{\sigma} & 0 & s(s^2 - 3\omega^2) \end{pmatrix},$$

where $q(s) = s^2(s^2 + \omega^2)$. Second, taking the inverse Laplace transform

$$e^{At} = \begin{pmatrix} 4 - 3\cos\omega t & \dfrac{\sin\omega t}{\omega} & 0 & 2\sigma\dfrac{1 - \cos\omega t}{\omega} \\ 3\omega\sin\omega t & \cos\omega t & 0 & 2\sigma\sin\omega t \\ \dfrac{6}{\sigma}(\sin\omega t - \omega t) & \dfrac{2\cos\omega t - 1}{\sigma} \,\dfrac{}{\omega} & 1 & -3t + \dfrac{4}{\omega}\sin\omega t \\ \dfrac{6}{\sigma}\omega(\cos\omega t - 1) & -\dfrac{2}{\sigma}\sin\omega t & 0 & 4\cos\omega t - 3 \end{pmatrix}.$$

1.2.5. Let A, B be 2 by 2 matrices with $a_{12} = b_{11} = -b_{22} = 1$ and all other entries equal zero.

1.2.6. Fix the matrix A and let $q_A(s)$ denote the eigenpolynomial of A. We show first that for f entire there is a polynomial $p(s)$, with coefficients depending *linearly* and *continuously* on f, and an entire function g such that

(*) $$f(s) = p(s) + q_A(s)g(s), \qquad s \text{ in } \mathbb{C}.$$

By "continuously" we mean that $|p^{(j)}(0)| \le k_j \max_C |f|$, where C is as in 1.2.1 and k_j is some positive constant depending only on j, C, and A. Let $q_j(s) = q_A(s)/(s - \lambda_j)^{m_j}, j = 1, \ldots, r$. Then $q_1(s), \ldots, q_r(s)$ are relatively prime which implies there are more polynomials $p_j(s)$ such that the sum of $p_j(s)q_j(s)$ over $1 \le j \le r$ equals 1 identically. Now let $r_j(s)$ be the Taylor expansion of f at λ_j up to order $m_j - 1$. Then, for each j, the coefficients of $r_j(s)$ depend linearly on f and by Cauchy's formula they also depend continuously on f. Moreover, by Taylor's theorem

(*$_j$) $$f(s) = r_j(s) + (s - \lambda_j)^{m_j}g_j(s), \qquad s \text{ in } \mathbb{C},$$

for some entire functions $g_j(s)$. Now $f(s)p_j(s)q_j(s)$ equals $r_j(s)p_j(s)q_j(s)$ plus $q_A(s)p_j(s)g_j(s)$ for all j, by (*$_j$). Summing over j, (*) follows. Since (*) implies that $f(A) = p(A)$, the solution is complete.

1.2.7. Choose c such that the eigenvalues of $\bar{A} = A - sI$ satisfy $\mathrm{Re}(\lambda) \le -k < 0$ for all s satisfying $\mathrm{Re}(s) \ge c$. Then, by 1.2.8, the Laplace transform

$$B(s) = \int_0^\infty e^{-st}e^{At}\, dt, \qquad \mathrm{Re}(s) \ge c,$$

is well defined. Now

$$(sI - A)B(s) = -\int_0^\infty (-se^{-st})e^{At} + e^{-st}(Ae^{At})\, dt$$

$$= -\int_0^\infty e^{At}d(e^{-st}) + e^{-st}d(e^{At})$$

$$= -\int_0^\infty d(e^{-st}e^{At}) = I.$$

1.2.8. Equation (2.4) shows that e^{tA} is a matrix whose entries are linear combinations of products of the form $p(t)e^{\lambda t}$ for some polynomials $p(t)$ and eigenvalues λ. Since $\text{Re}(\lambda) < -k$, we have $|p(t)e^{\lambda t}|e^{kt} \le |p(t)|e^{(\text{Re}(\lambda)+k)t}$ is bounded over $t \ge 0$. The result follows.

1.2.9. $\dot{x} = \bar{A}x$ iff $x(t) = e^{\bar{A}t}x(0)$, $t \ge 0$. Now $\bar{A}$ stable implies $x(t) \to 0$ as $t\uparrow\infty$ by 1.2.8. Conversely, suppose $Av = \lambda v$. Then $e^{t\lambda}v = e^{t\bar{A}}v = x(t) \to 0$ as $t\uparrow\infty$ which implies that $\text{Re}(\lambda) < 0$.

1.2.10. For $n \ge 1$ let $p_n(s; a_1, \ldots, a_n)$ be the determinant of

$$\begin{pmatrix} s & -1 & 0 & \cdots & 0 & 0 \\ 0 & s & -1 & \cdots & 0 & 0 \\ & & & \cdots & & \\ 0 & 0 & 0 & \cdots & s & -1 \\ a_1 & a_2 & a_3 & \cdots & a_{n-1} & a_n \end{pmatrix}.$$

Then, by expanding along the last column, we have

$$p_n(s; a_1, \ldots, a_n) = a_n s^{n-1} + p_{n-1}(s; a_1, \ldots, a_{n-1})$$

$$= \cdots = a_n s^{n-1} + a_{n-1}s^{n-2} + \cdots + a_1.$$

Since $\det(sI - \bar{A}) = p_n(s; q_1, q_2, \ldots, q_{n-1}, s + q_n) = q(s)$, we see that $q(s)$ is the eigenpolynomial of $\bar{A}$. The result follows.

1.2.11. Let K denote any solution of the Lyapunov equation and let $L(T)$denote the integral over $[0, T]$ of $e^{t\bar{A}^*}Qe^{t\bar{A}}$. Then by multiplying the Lyapunov equation on the left by $e^{t\bar{A}^*}$ and on the right by $e^{t\bar{A}}$, we have

$$L(T) = -\int_0^T e^{t\bar{A}^*}(\bar{A}^*K + K\bar{A})e^{t\bar{A}}\,dt = -\int_0^T d(e^{t\bar{A}^*}Ke^{t\bar{A}})$$

$$= K - e^{T\bar{A}^*}Ke^{T\bar{A}}.$$

Letting $T\uparrow\infty$, $L(T) \to L$ and $e^{T\bar{A}} \to 0$. Thus $K = L$.

1.2.12. Clearly, $Q = Q^*$ implies $L = L^*$. Also $Q \to L$ is onto: simply set $Q = -\bar{A}^*L - L\bar{A}$. Moreover, $L = 0$ implies $Q = 0$ so $Q \to L$ is a bijection. If $Q \ge 0$ then $v(t)^*Qv(t) \ge 0$ where $v(t) = e^{t\bar{A}}x$. Integrating over $t \ge 0$, we have $L \ge 0$.

1.2.13. Recall that here $|A|^2$ denotes the sum of the *squares* of the absolute values of the entries of A. We recall that the Cauchy–Schwartz inequality says that $|\text{trace}(AB^*)| \le |A||B|$ for any A and B. Let $f(T)$ and $g(T)$ denote the *squares* of the left-hand and right-hand sides of the stated inequality, respectively. Thus we need to show that $f(T) \le g(T)$. Since $f(0) = g(0) = 0$, this will follow as soon as we show that $f'(T) \le g'(T)$. But for any matrix-valued function of time $B(\cdot)$, the derivative of $|B(T)|^2$

equals $2 \, \mathrm{Re}(\mathrm{trace}(B(T)\dot{B}(T)^*))$ (check this!). Thus

$$f'(T) = 2 \, \mathrm{Re}\left(\mathrm{trace}\left(\left(\int_0^T A(t)\, dt\right)A(T)^*\right)\right)$$

$$= 2\int_0^T \mathrm{Re}(\mathrm{trace}(A(t)A(T)^*))\, dt$$

$$\le 2\int_0^T |A(t)|\,|A(T)|\, dt = 2\left(\int_0^T |A(t)|\, dt\right)|A(T)| = g'(T).$$

1.2.14. We use the fact that any positive matrix L can be written as $L = P^*P = P^2$ for some invertible *self-adjoint* $P = P^*$. Note also that $A \ge B$ holds iff $P^*AP \ge P^*BP$ holds for any invertible P; when P is also self-adjoint, this happens iff $PAP \ge PBP$ holds. We first prove the special case when $L_2 = I$. Then $L_1 = P_1^2$ and $L_1 \ge I$ imply $P_1 I P_1 \ge I$ which implies $I \ge P_1^{-1}P_1^{-1} = L_1^{-1}$. For the general case write $L_2 = P_2^2$. Then $L_1 \ge L_2$ implies $P_2^{-1}L_1 P_2^{-1} \ge I$ which by the special case implies $I \ge (P_2^{-1}L_1 P_2^{-1})^{-1} = P_2 L_1^{-1} P_2$ which implies $L_2^{-1} = P_2^{-1}IP_2^{-1} \ge P_2^{-1}P_2 L_1^{-1}P_2 P_2^{-1} = L_1^{-1}$. For the second statement note that the trace is a linear function of a matrix and that $A \ge 0$ implies all eigenvalues are ≥ 0 which implies $\mathrm{trace}(A) \ge 0$.

1.3.1.

$$\frac{d}{dt}\left(\int_0^t e^{(t-s)A}Bu(s)\, ds\right) = \int_0^t Ae^{(t-s)A}Bu(s)\, ds + Bu(t)$$

$$= Ax^u(t) - Ae^{tA}x^0 + Bu(t)$$

$$= Ax^u(t) + Bu(t) - \frac{d}{dt}(e^{tA}x^0).$$

Also if $x_1(\cdot)$, $x_2(\cdot)$ are solutions then $x(\cdot) = x_1(\cdot) - x_2(\cdot)$ satisfies $\dot{x} = Ax$, $x(0) = 0$ which implies $x(t) = e^{tA}x(0) = 0$, $t \ge 0$.

1.3.2. A is given by the matrix appearing in 1.2.10 while $b = (0, 0, \ldots, 0, 1)^*$ and $c = (p_1, p_2, \ldots, p_n)$. Also by 1.2.10, the eigenpolynomial of A is $q(s) = s^n + q_n s^{n-1} + \cdots + q_1$. Since $(sI - A)$ times $\mathrm{col}(1, s, \ldots, s^{n-1})$ equals $q(s)b$, we have

$$g(s) = c(sI - A)^{-1}b = (p_1, \ldots, p_n)\frac{1}{q(s)}\begin{pmatrix} 1 \\ s \\ \vdots \\ s^{n-1} \end{pmatrix} = \frac{p(s)}{q(s)},$$

with $p(s) = p_n s^{n-1} + \cdots + p_1$.

1.3.3. Both sides of the equation satisfy the same differential equation and both sides equal $x^u(T; x^0)$ when $t = 0$. By uniqueness, they are equal.

1.3.4. Referring to the solution of 1.2.4,

$$G(s) = C(sI - A)^{-1}B = \frac{1}{m\sigma q(s)}\begin{pmatrix} \sigma s^2 & 2\omega\sigma s \\ -2\omega s & s^2 - 3\omega^2 \end{pmatrix}.$$

1.3.5. Note that since $|e^{(t-s)\bar{A}}B| \le \mathrm{const} = k$ for $t \ge s$, we have

(E) $|x^u(t; x^0)| \le |e^{t\bar{A}}x^0| + k\int_0^t |u(s)|\, ds \le |e^{t\bar{A}}x^0| + k\int_0^\infty |u(s)|\, ds.$

For $T > 0$ let $v(t) = u(t + T)$, $t \geq 0$. Then $u *_T v = u$ and $x^v(t - T; x^u(T)) = x^u(t)$, $t \geq T$, by 1.3.3. Applying the above inequality yields

$$|x^u(t)| = |x^v(t - T; x^u(T))|$$

$$\leq |e^{(t-T)\bar{A}} x^u(T)| + k \int_0^\infty |v(s)| \, ds, \qquad t \geq T.$$

Thus

$$\lim_{t \uparrow \infty} |x^u(t)| \leq k \int_0^\infty |v(s)| \, ds = k \int_T^\infty |u(s)| \, ds.$$

Since T is arbitrary, letting $T \uparrow \infty$ the result follows.

1.3.6. Choose $\varepsilon > 0$ such that $\bar{A}_\varepsilon = \bar{A} + \varepsilon I$ is stable and let $L > 0$ be the unique solution of $\bar{A}_\varepsilon^* L + L \bar{A}_\varepsilon + I = 0$. Set $b = |B^* LLB|$ and $\bar{a} = -2\varepsilon$. Then

$$\frac{d}{dt} x^* L x = x^* (\bar{A}^* L + L \bar{A}) x + 2 u^* B^* L x$$

$$\leq -2\varepsilon x^* L x - |x|^2 + |u^* B^* L|^2 + |x|^2$$

$$\leq \bar{a} x^* L x + b |u|^2.$$

Thus with $s(\cdot)$ as defined, $\delta(t) = s(t) - x(t)^* L x(t)$ satisfies

$$\dot{\delta} = \dot{s} - (x^* L x)^{\cdot} = \bar{a} s + b |u|^2 - (x^* L x)^{\cdot}$$

$$\geq \bar{a} s + b |u|^2 - \bar{a}(x^* L x) - b |u|^2 = \bar{a} \delta.$$

Hence $\delta(t) \geq e^{\bar{a} t} \delta(0) = 0$. Applying 1.3.5 to $s(\cdot)$, the result follows.

1.3.7. With notation as in the problem statement, set

$$k = \left(\int_0^\infty |e^{\bar{A} t} B|^{r'} \, dt \right)^{1/r'} < +\infty.$$

Then using Holder's inequality

$$|x^u(t; x^0)| \leq |e^{t\bar{A}} x^0| + \int_0^t |e^{(t-s)\bar{A}} B| |u(s)| \, ds$$

$$\leq |e^{t\bar{A}} x^0| + \left(\int_0^t |e^{(t-s)\bar{A}} B|^{r'} \, ds \right)^{1/r'} \left(\int_0^t |u(s)|^r \, ds \right)^{1/r}$$

and so

$$(E_r) \qquad |x^u(t; x^0)| \leq |e^{t\bar{A}} x^0| + k \left(\int_0^\infty |u(s)|^r \, ds \right)^{1/r}.$$

The solution is completed by continuing as in the proof of 1.3.5 with inequality (E_r) playing the role of inequality (E).

1.3.8. Young's inequality states that for $k(\cdot)$ satisfying (3.5) and $u(\cdot)$ satisfying (3.6) the *convolution*

$$w(t) = (k * u)(t) = \int_0^t k(t - s) u(s) \, ds, \qquad t \geq 0,$$

satisfies (3.6) and

$$\int_0^\infty |w(t)|^r \, dt \le c^r \int_0^\infty |u(t)|^r \, dt, \qquad c = \int_0^\infty |k(t)| \, dt.$$

Set $k(t) = e^{\bar{A}t}$ and $u(\cdot) = Bv(\cdot)$. Then $x^v(\cdot) = k(\cdot)x^0 + w(\cdot)$. Since $k(\cdot)$ also satisfies (3.6), we have

$$\int_0^\infty |x^v(t)|^r \, dt = \int_0^\infty |k(t)x^0 + w(t)|^r \, dt$$

$$\le 2^r \left(|x^0|^r \int_0^\infty |k(t)|^r \, dt + \int_0^\infty |w(t)|^r \, dt \right).$$

The result follows.

1.4.2. Let e_j denote the jth standard basis vector. Then $b_1 = e_2/m$ and $b_2 = e_4/m\sigma$. Now

$$(e_2, Ae_2, A^2 e_2, A^3 e_2) = \begin{pmatrix} 0 & 1 & 0 & -\omega^2 \\ 1 & 0 & -\omega^2 & 0 \\ 0 & 0 & -2\dfrac{\omega}{\sigma} & 0 \\ 0 & -\dfrac{2}{\sigma}\omega & 0 & 2\dfrac{\omega^3}{\sigma} \end{pmatrix},$$

$$(e_4, Ae_4, A^2 e_4, A^3 e_4) = \begin{pmatrix} 0 & 0 & 2\sigma\omega & 0 \\ 0 & 2\sigma\omega & 0 & -2\omega^3\sigma \\ 0 & 1 & 0 & -4\omega^2 \\ 1 & 0 & -4\omega^2 & 0 \end{pmatrix}.$$

Thus $\det(b_1, Ab_1, A^2 b_1, A^3 b_1) = 0$ and so (A, b_1) is not a controllable pair. Now $\det(b_2, Ab_2, A^2 b_2, A^3 b_2) = -12\omega^4/m^4\sigma^2 \ne 0$. Thus (A, b_2) is controllable and hence (A, B) is also controllable.

1.4.3. Let $q(s) = s^n + q_n s^{n-1} + \cdots + q_1$ be the eigenpolynomial of A_1. Let $v_j = A_1^{n-j} b_1 + q_n A_1^{n-j-1} b_1 + \cdots + q_{j+1} b_1$, $j = 1, \ldots, n-1$, and $v_n = b_1$. Then because (A_1, b_1) is controllable, $\{v_1, \ldots, v_n\}$ forms a basis for $\mathbb{C}^n$. Set $p_j = c_1 v_j, j = 1, \ldots, n$. Then

$$A_1 v_j = v_{j-1} - q_j v_n, \qquad 2 \le j \le n,$$

$$A_1 v_1 = A_1^n b_1 + q_n A_1^{n-1} b_1 + \cdots + q_2 A_1 b_1 = q(A_1)b_1 - q_1 b_1 = -q_1 v_n.$$

Thus with $P = [v_1, \ldots, v_n]$, $A = P^{-1} A_1 P$, $b = P^{-1} b_1$, and $c = c_1 P$ are as in 1.3.2. Since the transfer function of (A_1, b_1, c_1) is equal to that of (A, b, c), we see, by 1.3.2, that the transfer function of (A_1, b_1, c_1) is $p(s)/q(s)$.

1.4.4. x is in V_T iff for some $u(\cdot)$ $x^u(T; x) = 0$. Since $x^u(T; x) = e^{TA}x + x^u(T; 0)$, this happens iff $e^{TA}x = x^{-u}(T; 0)$ which happens iff $e^{TA}x$ is in V^T. Since e^{AT} is an invertible matrix, $V_T = \mathbb{C}^n$ iff $V^T = \mathbb{C}^n$.

1.4.6. $\mathrm{I/O}(x_1^0) = \mathrm{I/O}(x_2^0)$ iff $x_1^0 - x_2^0$ is in W^T for all $T > 0$. Therefore observability implies $x_1^0 = x_2^0$. Conversely, lack of observability implies there is an $x \ne 0$ in W^T and so $\mathrm{I/O}(x) = \mathrm{I/O}(0)$.

1.4.7. Look at the proof of 1.4.5.

1.4.8. Suppose $x^*Lx = 0$. Since

$$x^*Lx = \int_0^\infty x^* e^{t\bar{A}^*} Q e^{t\bar{A}} x \, dt = \int_0^\infty |Ce^{t\bar{A}}x|^2 \, dt,$$

this implies that $Ce^{t\bar{A}}x = 0$, $t \geq 0$. By 1.4.7, $x = 0$. Thus $L > 0$.

1.4.9. Let $\bar{A} = A - BF$. It is enough to show that (A, B) controllable implies $(\bar{A}, B)$ controllable. Indeed, applying this result with $-F$ replacing F, the converse follows.

First method: The columns of $(B, AB, \ldots, A^{n-1}B)$ are linear combinations of the columns of $(B, \bar{A}B, \ldots, \bar{A}^{n-1}B)$. Thus if the rank of the first matrix is n then so is the rank of the second.

Second method: Let $x^u(\cdot)$, V^T, correspond to (A, B) and let $\bar{x}^u(\cdot)$, $\bar{V}^T$, correspond to $(\bar{A}, B)$. Given $u(\cdot)$ set $v(\cdot) = u(\cdot) + Fx^u(\cdot; 0)$. Since $Ax^u + Bu = \bar{A}x^u + B(u + Fx) = \bar{A}x + Bv$, we have $x^u(\cdot) = \bar{x}^v(\cdot)$. Thus $V^T \subset \bar{V}^T$.

1.4.10. (A, B) is controllable iff $(B, AB, \ldots, A^{n-1}B)$ has rank n iff

$$\begin{pmatrix} B^* \\ B^*A^* \\ \vdots \\ B^*A^{*n-1} \end{pmatrix}$$

has rank n iff (A^*, B^*) is observable.

1.4.11. Set $G_j(s) = C_j(sI - A_j)^{-1}B_j$, $j = 1, 2$. Equivalence means $G_1(s) = G_2(s)$ which implies $G_1(s^*)^* = G_2(s^*)^*$ which implies the result.

1.4.12. This implies that $G(s) = G(s^*)^*$, i.e., that $G(s)$ is symmetric. It can then be shown that there is an equivalent symmetric triple (A', B', C') i.e., $A'^* = A'$ and $B'^* = C'$.

1.4.13. Suppose $m(A) = 0$ for some polynomial $m(s)$ of degree r less than n. Then multiplying by b, we have a linear relation among $b, Ab, \ldots, A^{n-1}b$. But controllability says that these vectors are linearly independent, since $\det(b, Ab, \ldots, A^{n-1}b) \neq 0$. Thus there is no polynomial $m(s)$ of degree smaller than n that annihilates A.

1.5.4. Clearly (A, B, C) Hamiltonian implies $-A^* = JAJ^{-1}$, $JC^* = B$, $-B^*J^{-1} = C$. This yields $G(-s^*)^* = (C(-s^*I - A)^{-1}B)^* = -B^*(sI + A^*)^{-1}C^* = -B^*(sI - JAJ^{-1})^{-1}C^* = -B^*J^{-1}(sI - A)^{-1}JC^* = C(sI - A)^{-1}B = G(s)$.

1.5.5. Let $m(s)$ denote the minimal polynomial of A. Clearly, $m(s)G(s) = Cm(s)(sI - A)^{-1}B$ then has polynomial entries. Since (A, B, C) is minimal, we can choose n by p matrices $F_1, \ldots, F_n$ and m by n matrices $H_1, \ldots, H_n$ such that

$$\sum_{i=1}^n F_i CA^{i-1} = I_n = \sum_{j=1}^n A^{j-1}BH_j.$$

By induction one checks that $CA^i(sI - A)^{-1}A^jB = s^{i+j}G(s) + E_{ij}(s)$ for some polynomial $E_{ij}(s)$. Setting $F(s) = F_1 + F_2 s + \cdots + F_n s^{n-1}$, $H(s) = H_1 + H_2 s + \cdots + H_n s^{n-1}$, we have

$$(sI - A)^{-1} = \sum_{i,j=1}^n F_i CA^{i-1}(sI - A)^{-1}A^{j-1}BH_j$$

$$= \sum_{i,j=1}^n s^{i-1}s^{j-1}F_i G(s)H_j + E(s)$$

$$= F(s)G(s)H(s) + E(s)$$

for some polynomial matrix $E(s)$. Multiplying this last equation by $p(s)$, we see that $p(s)(sI - A)^{-1}$ has polynomial entries as soon as $p(s)G(s)$ has. Thus $m(s)$ divides $p(s)$, which implies that the denominator of $G(s)$ is $m(s)$.

1.5.6. We are given that (A, B) is controllable. Then, as in 1.5.5,

$$C(sI - A)^{-1} = G(s)H(s) + E(s)$$

for some polynomial matrices $H(s)$ and $E(s)$. Suppose that $Ax = \lambda x$ with $Cx \neq 0$. Applying x to both sides of this last equation yields

$$\frac{Cx}{s - \lambda} = G(s)H(s)x + E(s)x.$$

Since $Cx \neq 0$, as $s \to \lambda$ the left-hand side blows up, and thus the right-hand side blows up. Thus λ must be a pole of $G(s)$.

1.5.7. If λ is unobservable then $Ax = \lambda x$ with $Cx = 0$ and $x \neq 0$. This implies that x is in W^T which implies unobservability of (A, C). Conversely, suppose that (A, C) is not observable. Then W^T is a nontrivial subspace with $A(W^T) \subset W^T$. Thus $A|_W$ must have a nonzero eigenvector x and an eigenvalue λ. Since x is in W^T by definition, λ is unobservable.

1.5.8. $q(s) = \det(sI - A) = \det(J)^2 \det(sI - A) = \det(J) \det(sJ - JA) = \det(J) \det(sJ + A^*J) = \det(J) \det(sI + A^*) \det(J) = \det(sI + A^*) = \det(-sI - A^*) = (\det(-s^*I - A))^* = q(-s^*)^*$. Thus λ is a root of $q(s)$ iff $-\lambda^*$ is a root of $q(s)$. If (A, B) is controllable, then there exist polynomial matrices $H(s)$ and $E(s)$ with $C(sI - A)^{-1} = G(s)H(s) + E(s)$. Thus there exist polynomial matrices $F(s)$ and $E'(s)$ with $(C(-s^* - A)^{-1})^* = F(s)G(-s^*)^* + E'(s) = F(s)G(s) + E'(s)$ which implies $(sI - A)^{-1}B = -JF(s)G(s) - JE'(s)$. By 1.5.7, (A, C) is observable.

1.6.3.

$$(B, AB, \ldots, A^{n-1}B) = \begin{pmatrix} 0 & 0 & \ldots & 0 & I \\ 0 & 0 & \ldots & I & * \\ \vdots & \vdots & & \vdots & \vdots \\ 0 & I & \ldots & * & * \\ I & * & \ldots & * & * \end{pmatrix}.$$

1.6.4. If the standard controllable realization were not minimal, then there would exist a realization with dimension less than $mr = r$. Thus the denominator of $G(s)$ would have degree less than r, which contradicts the definition of r.

1.6.6. By 1.3.4, $C(sI - A)^{-1}b_1$ is given by $\begin{pmatrix} \sigma s \\ -2\omega \end{pmatrix} / m\sigma s(s^2 + \omega^2)$. Since $m = 1$ here, the standard controllable realization is minimal: here the denominator equals $s^3 + s\omega^2$, $p = 2$, $r = 3$, $q_3 = 0$, $q_2 = \omega^2$, $q_1 = 0$, $P_3 = 0$, $P_2 = \begin{pmatrix} 1/m \\ 0 \end{pmatrix}$, and $P_1 = -2\omega \begin{pmatrix} 0 \\ 1 \end{pmatrix} / m\sigma$. Thus

$$A = \begin{pmatrix} 0 & 1 & 0 \\ 0 & 0 & 1 \\ 0 & -\omega^2 & 0 \end{pmatrix}, \quad b = \begin{pmatrix} 0 \\ 0 \\ 1 \end{pmatrix}, \quad C = \begin{pmatrix} 0 & \dfrac{1}{m} & 0 \\ \dfrac{-2\omega}{\sigma m} & 0 & 0 \end{pmatrix}.$$

1.6.7. Choose (A, B, C) such (A^*, C^*, B^*) is the standard controllable realization of $G(s^*)^*$.

1.6.9. Since $\text{rank}(B) = \text{rank}(B_1) + \cdots + \text{rank}(B_N) = n_1 + \cdots + n_N$, (A, B) is controllable. Now suppose $x = \text{col}(x_1, \ldots, x_n)$ satisfies $Ce^{tA}x = 0$ for $t \geq 0$. Then since A is diagonal

$$e^{t\lambda_1}C_1 x_1 + \cdots + e^{t\lambda_N}C_N x_N = 0 \qquad \text{for} \quad t \geq 0.$$

Since the λ's are distinct, this implies $C_1 x_1 = \cdots = C_N x_N = 0$. Since $\text{rank}(C_j) = n_j$ we must have $x_j = 0$ which implies $x = 0$.

Chapter 2

2.1.2. No. Take A stable and $B = 0$.

2.1.4. By 1.4.3 there exists P invertible with $P^{-1}AP$, $P^{-1}b$ in standard controllable form. By the discussion in Section 2.1 there is an f_1 with $\det(sI - \bar{A}_1)$, $\bar{A}_1 = P^{-1}AP - P^{-1}bf_1$, having prescribed eigenvalues. Since $A - bf$, $f = f_1 P^{-1}$, has the same eigenpolynomial as $\bar{A}_1$, the result follows.

2.1.6. Since $cb \neq 0$, there is a unique positive solution $k > 0$ to the displayed quadratic equation, $k = (-\text{Re}(a) + \sqrt{\text{Re}(a)^2 + |cb|^2})/|b|^2$. Now suppose that l is any solution of the Lyapunov equation. Then by subtracting the two equations we see that $l - k$ satisfies

$$\bar{a}^*(l - k) + (l - k)\bar{a} + |f - b^*k|^2 = 0, \qquad \bar{a} = a - bf.$$

Since this also is a Lyapunov equation with $q = |f - b^*k|^2 \geq 0$, we see that $l - k \geq 0$ (1.2.12) with equality iff $f - b^*k = 0$.

2.1.7. Let $x(\cdot)$, $y(\cdot)$ denote the state and output trajectories starting from zero, corresponding to a given $u(\cdot)$. Let $\hat{u}(s)$, $\hat{x}(s)$, and $\hat{y}(s)$ denote their Laplace transforms. Then $\hat{y}(s) = C\hat{x}(s)$, $\hat{x}(s) = (sI - A)^{-1}B\hat{u}(s)$, and $\hat{y}(s) = G(s)\hat{u}(s)$. Now $(sI - \bar{A})\hat{x}(s) = (sI - A)\hat{x}(s) + BF\hat{y}(s)$ with $\bar{A} = A - BFC$. Thus $(sI - \bar{A})\hat{x}(s) = B\hat{u}(s) + BF\hat{y}(s) = B(I + FG(s))\hat{u}(s)$. Hence $G(s)\hat{u}(s) = C\hat{x}(s) = C(sI - \bar{A})^{-1}B(I + FG(s))\hat{u}(s) = G_F(s)(I + FG(s))\hat{u}(s)$. We thus obtain

$$G(s) = G_F(s)(I + FG(s)).$$

In case of scalar inputs and outputs, $m = p = 1$, this reduces to

$$(*) \qquad\qquad g_f(s) = \frac{g(s)}{1 + fg(s)}.$$

2.1.8. Assume that (A, b, c) is minimal. Then, by 1.4.13 and 1.5.5, the denominator of $g(s) = c(sI - A)^{-1}b$ is $q(s) = \det(sI - A)$. Set $p(s) = g(s)q(s)$. Now, by 2.1.7, $(\bar{A}, b, c)$ is also minimal. Similarly, it follows that the denominator of $g_1(s) = c(sI - \bar{A})^{-1}b$, $\bar{A} = A - bc$, is $\det(sI - A + bc)$. On the other hand, the denominator of $g(s)/(1 + g(s)) = p(s)/(p(s) + q(s))$ is $p(s) + q(s)$ as $p(s)$ and $q(s)$ have no common factors. By $(*)$ above, these last two denominators must agree and so, for s not in $\text{spec}(A)$,

$$\det(sI - A + bc) = p(s) + q(s) = \det(sI - A)(1 + c(sI - A)^{-1}b).$$

Now both sides of this equation, for s fixed, are continuous functions of the entries of A, b, and c. Since any triple (A, b, c) can be approximated arbitrarily closed by one that is minimal, the equation is valid for all triples. Taking $A = 0$ and $s = 1$, the special case follows.

2.2.3. $S(0) = 0$, $S(x) = +\infty$ when $x \neq 0$.

2.2.5. Choose $v(\cdot)$ such that $x^v(T; x^0) = 0$ and let $u(\cdot) = (v *_T 0)(\cdot)$ (1.3.3). Then $x^u(t; x^0) = 0$ for $t \geq T$ and so $J^u(x^0) < \infty$ which implies $S(x^0) < \infty$.

2.2.6. $S(x) = 0$ iff $J^u(x) = 0$ for some $u(\cdot)$ iff $u(\cdot) = Cx^u(\cdot) = 0$ iff $Ce^{tA}x = 0$, $t \geq 0$, iff x is in W^T.

2.2.7. $\sin(x)$ is not proper, take $x_k = k\pi$; $|x|^2$ is proper, e^{x^2} is not proper on the complexes but is proper on the reals.

2.3.1. With $F = F_0 + \varepsilon D$, D arbitrary, we have

$$\bar{A}_\varepsilon^* L_\varepsilon + L_\varepsilon \bar{A}_\varepsilon + F_\varepsilon^* F_\varepsilon + C^* C = 0, \qquad \bar{A}_\varepsilon = A - BF_\varepsilon.$$

Since $L_\varepsilon \geq L_0 = K$, differentiation with respect to ε at $\varepsilon = 0$ yields $dL_\varepsilon/d\varepsilon = 0$. Thus $(-BD)^* K + K(-BD) + D^* F_0 + F_0 D = 0$ which yields $Q + Q^* = 0$, $Q = D^*(F_0 - B^* K)$. Replacing D by iD, $i = \sqrt{-1}$, we have $Q - Q^* = 0$ and hence $Q = 0$.

2.3.2. Taking the adjoint of (3.1), we see that K^* also satisfies (3.1). By uniqueness, $K = K^*$.

2.3.3. If $K = K^*$ satisfies (3.1) with $\bar{A} = A - BB^* K$ stable, then $\bar{A}^* K + K\bar{A} + Q = 0$, where $Q = KBB^* K + C^* C$. Since $Q \geq 0$, it follows that $K \geq 0$. Also since $Q \geq C^* C$ and $(\bar{A}, C)$ is observable, it follows that $K > 0$ (1.4.8).

2.3.9. Using the Riccati equation,

$$
\begin{aligned}
G(-s^*)^* G(s) + I &= B^*(-sI - A^*)^{-1} C^* C(sI - A)^{-1} B + I \\
&= B^*(-sI - A^*)^{-1}(-A^* K - KA + KBB^* K)(sI - A)^{-1} B + I \\
&= B^*(-sI - A^*)^{-1}((-sI - A^*)K + K(sI - A) \\
&\quad + KBB^* K)(sI - A)^{-1} B + I \\
&= B^* K(sI - A)^{-1} B + B^*(-sI - A^*)^{-1} KB \\
&\quad + B^*(-sI - A^*)^{-1} KBB^* K(sI - A)^{-1} B + I \\
&= G^\#(s) + G^\#(-s^*)^* + G^\#(-s^*)^* G^\#(s) + I \\
&= (I + G^\#(-s^*))^*(I + G^\#(s)).
\end{aligned}
$$

2.3.10. See 2.3.9.

2.3.11. Let $u(\cdot)$, $x(\cdot)$, and $y(\cdot)$ be the input, state, and output trajectories starting from zero corresponding to the transfer function $I + G^\#(s)$. Let $\hat{u}(s)$, $\hat{x}(s)$, and $\hat{y}(s)$ denote their Laplace transforms. Then $(sI - A)\hat{x}(s) = \hat{u}(s)$, $\hat{y}(s) = B^* K\hat{x}(s) + \hat{u}(s)$, and $\hat{y}(s) = (I + G^\#(s))\hat{u}(s)$. To compute the inverse we have to express $\hat{u}(s)$ in terms of $\hat{y}(s)$. Now $\hat{u}(s) = \hat{y}(s) - B^* K\hat{x}(s)$ and $(sI - \bar{A})\hat{x}(s) = (sI - A)\hat{x}(s) + B(\hat{y}(s) - \hat{u}(s)) = \hat{y}(s)$ so $\hat{u}(s) = \hat{y}(s) - B^* K(sI - \bar{A})^{-1}\hat{y}(s)$; thus $\hat{u}(s) = (I - \bar{G}^\#(s))\hat{y}(s)$.

2.3.12. First, note that

$$\tilde{A}_1 = \tilde{A} + \tilde{B}\tilde{C} = \begin{pmatrix} A & 0 \\ C^*C & -A^* \end{pmatrix}.$$

Second, by induction check that

$$(\tilde{A} + \tilde{B}\tilde{C})^{k+1}\tilde{B} = \begin{pmatrix} A^{k+1}B \\ \sum_{i+j=k}(-1)^i A^{*i}C^*CA^jB \end{pmatrix}, \qquad k \geq 0.$$

Thus $\tilde{G}_{-1}(s) = \tilde{C}(sI - \tilde{A} - \tilde{B}\tilde{C})^{-1}\tilde{B}$ equals

$$\tilde{G}_{-1}(s) = \sum_{k \geq 0} \tilde{C}\tilde{A}_1^{k+1}\tilde{B}\frac{1}{s^{k+2}} = \sum_{k \geq 0}\sum_{i+j=k}(-1)^{i+1}\frac{B^*A^{*i}C^*}{s^{i+1}}\frac{CA^jB}{s^{j+1}}$$

$$= \left(\sum_{i \geq 0}\frac{CA^iB}{(-s^*)^{i+1}}\right)^*\left(\sum_{j \geq 0}\frac{CA^jB}{s^{j+1}}\right) = G(-s^*)^*G(s).$$

Then, by 2.1.7,

$$\tilde{G}(s) = \tilde{G}_{-1}(s)(I + \tilde{G}_{-1}(s))^{-1} = G(-s^*)^*G(s)(I + G(-s^*)^*G(s))^{-1}.$$

In the scalar case $m = p = 1$, this reduces to

$$\tilde{g}(s) = \frac{g(-s^*)^*g(s)}{1 + g(-s^*)^*g(s)}.$$

2.3.13. That $(\tilde{A}, \tilde{B}, \tilde{C})$ is Hamiltonian follows from the definition. Also

$$\tilde{A}\tilde{x} = \begin{pmatrix} A & BB^* \\ C^*C & -A^* \end{pmatrix}\begin{pmatrix} x \\ -Kx \end{pmatrix} = \begin{pmatrix} Ax - BB^*Kx \\ C^*Cx + A^*Kx \end{pmatrix} = \begin{pmatrix} \bar{A}x \\ -K\bar{A}x \end{pmatrix} = (\bar{A}x)^\sim,$$

using the Riccati equation. Set $\tilde{q}(s) = \det(sI - A)$. Since $\tilde{A}|_V = \bar{A}$ we have $\det(sI - \bar{A})$ divides $\tilde{q}(s)$ (1.2.3) and hence $\det(sI + \bar{A}^*)$ divides $\tilde{q}(-s^*)^* = \tilde{q}(s)$ (1.5.8). Since these two have no common factors, their product is a monic polynomial of degree $2n$ dividing $\tilde{q}(s)$. The result follows.

2.3.14. Assume that $(\tilde{A}, \tilde{B}, \tilde{C})$ is minimal. If $CA^ix = 0$ for all i, then $\tilde{C}\tilde{A}^i\begin{pmatrix} x \\ 0 \end{pmatrix} = 0$ for all i. This shows that (A, C) is observable. Similarly, if c is a row vector with $cA^iB = 0$ for all i then $(c, 0)\tilde{A}^i\tilde{B} = 0$ for all i which shows that (A, B) is controllable. Finally, suppose that $B^*KA^ix = 0$ for all i. Then, by 2.3.13, $\tilde{C}\tilde{A}^i\tilde{x} = 0$ for all i. Thus (A, B^*K) is also observable.

2.3.15. Here $m = p = 1$, $n = 2$. Thus

$$A = \begin{pmatrix} 0 & 1 \\ 0 & 1 \end{pmatrix}, \qquad b = \begin{pmatrix} 0 \\ 1 \end{pmatrix}, \qquad c = (1, 1)$$

is a minimal realization. Setting

$$K = \begin{pmatrix} a & b \\ b & c \end{pmatrix},$$

the Riccati equation reduces to $0 = b^2 - 1, 2(b + c) = c^2 - 1, a + b = bc - 1$. Solving yields

$$K = \begin{pmatrix} 1 & 1 \\ 1 & 3 \end{pmatrix}, \qquad b^*K = (1, 3), \qquad \bar{A} = A - bb^*K = \begin{pmatrix} 0 & 1 \\ -1 & -2 \end{pmatrix}.$$

Thus

$$g^\#(s) = \frac{3s + 1}{s^2 - s}, \quad \bar{g}^\#(s) = \frac{g^\#(s)}{1 + g^\#(s)} = \frac{3s + 1}{(s + 1)^2}, \quad \bar{g}(s) = \frac{1}{s + 1}, \text{ and } \tilde{g}(s) = \frac{1}{1 - s^2}.$$

2.4.2. To show uniqueness it is enough to show that if $x(t)$, $0 \le t \le T$, satisfies $\dot{x} = A(t)x$ and $x(0) = 0$, then $x(t) = 0, 0 \le t \le T$. Let $a(\cdot)$ be as in the proof of 2.4.1 and set

$$\mu(t) = \max_{0 \le s \le t} e^{-a(s)} |x(s)|.$$

Then $x(t) = \int_0^t A(s)x(s)\, ds$ implies

$$e^{-a(t)} |x(t)| \le e^{-a(t)} \int_0^t |A(s)| |x(s)|\, ds \le \left(\int_0^t e^{a(s)} \dot{a}(s) e^{-a(t)}\, ds \right) \mu(t)$$

$$= (1 - e^{-a(t)}) \mu(t) \le (1 - e^{-a(T)}) \mu(T).$$

Taking the maximum of the left-hand side over $0 \le t \le T$, we end up with $\mu(T) \le (1 - e^{-a(T)}) \mu(T)$. If $a(T) = 0$, there is nothing to prove. If $a(T) > 0$, then this last inequality implies that $\mu(T) = 0$; the result follows.

2.4.8. Equation (4.3) implies $x^*M(1)x \le \int_0^1 |u(t)|^2 + |Cx^v(t; x)|^2\, dt$ for all x. Let $c = c(A, B, C)$ be the inverse of the smallest eigenvalue of $M(1)$. Then by choosing $v(\cdot) = u(\cdot + T)$ and $x = x^u(T)$, we have

$$|x^u(T)|^2 \le cx^u(T)^*M(1)x^u(T) \le c \int_0^1 |u(t + T)|^2 + |Cx^v(t; x^u(T))|^2\, dt$$

$$\le c \int_T^{T+1} |u(t)|^2 + |Cx^u(t)|^2\, dt.$$

Integrating over $T \ge 0$ and interchanging the order of integration, the result follows.

2.5.1. Given (A, B, C) let M denote any of the $\binom{nm}{n}$ n by n submatrices of $(B, AB, \ldots, A^{n-1}B)$ and set

$$h(A, B) = \sum |\det(M)|^2,$$

where the sum is over all such M. Then (A, B) is a controllable pair iff $h(A, B) \ne 0$. Similarly, set $f(A, C) = h(A^*, C^*)^*$. Then (A, C) is an observable pair iff $f(A, C) \ne 0$. Moreover, (A, B, C) is minimal iff $f(A, C)h(A, B) \ne 0$. Since fh is a polynomial, it is continuous, thus any triple sufficiently close to a triple satisfying $fh \ne 0$ also satisfies $fh \ne 0$; moreover, for any triple satisfying $fh = 0$, there is a triple that is minimal and arbitrarily close to the given triple; otherwise fh would vanish on an open set, which is impossible for a nonzero polynomial.

2.5.2. fh is an example of such a polynomial.

2.5.3. This follows from $\det(AB) = \det(A)\det(B)$ and $\det(A) \ne 0$ iff A is invertible.

2.5.6. Equations (5.3) and (5.4) imply $K(A, B, C) = J^*K(JAJ^{-1}, JB, CJ^{-1})J = J^*K(-A^*, -C^*, B^*)J = J^*K(A, B, C)^{-1}J$. Thus $JK = JJ^*K^{-1}J = K^{-1}J$. Also when $A^* = -A, B^* = C$, (5.4) implies $K = K^{-1}$.

2.5.9. Equation (5.6) says that $K(\alpha)$ solves

$$A^*K + KA = KBB^*K - \alpha C^*C, \qquad A = A(\alpha). \qquad (*)$$

Applying 2.3.10 with $s = i\omega$ and $(A(\alpha), B, \sqrt{\alpha}C)$ we have

$$(I - \bar{G}_\alpha^\#(i\omega))^*(I - \bar{G}_\alpha^\#(i\omega)) = I - \alpha\bar{G}(i\omega)^*\bar{G}(i\omega)$$

$$\geq (1 - \alpha)I$$

by bounded realness.

2.5.10. Since $(\bar{A}, C)$ is observable this follows from 2.3.3.

2.5.11. Suppose that λ is *not* a pole of $G_\alpha^\#(s)$. Then, by 1.5.6, $B^*K(\alpha)x = 0$, $A(\alpha)x = \lambda x$ for some nonzero x. Thus $\bar{A}x = \lambda x$ and $x^*Kx > 0$. Multiplying $(*)$ above by x on the left as well as on the right,

$$(\lambda + \lambda^*)x^*Kx = 0 - \alpha|Cx|^2.$$

Thus $\mathrm{Re}(\lambda) \leq 0$. Since $(\bar{A}, C)$ is observable, $Cx \neq 0$, and so $\mathrm{Re}(\lambda) < 0$.

2.5.12. The first statement follows from 2.3.11. Now

$$|I + G_\alpha^\#(i\omega)|^2 = \mathrm{trace}((I + G_\alpha^\#(i\omega))^*(I + G_\alpha^\#(i\omega)))$$

$$\leq \mathrm{trace}\left(\frac{I}{1-\alpha}\right) = \frac{n}{1-\alpha},$$

by 2.5.9.

2.5.14. Multiply (5.6) on both sides by K^{-1}. This yields

$$\bar{A}K^{-1} + K^{-1}\bar{A}^* + Q = 0, \qquad Q = BB^* + K^{-1}C^*CK^{-1},$$

$$\bar{A}L + L\bar{A}^* + Q' = 0, \qquad Q' = BB^*.$$

Since $Q \geq Q'$, 1.2.12 implies $K^{-1} \geq L$ which implies, by 1.2.14, that $K \leq L^{-1}$.

2.5.15. By definition of f, (5.9) implies

$$0 = A^*K'(\alpha) + K'(\alpha)A + C^*C$$

$$= \bar{A}^*K'(\alpha) + K'(\alpha)\bar{A} + K(\alpha)BB^*K'(\alpha) + K'(\alpha)BB^*K(\alpha) + C^*C, \qquad K'(\alpha) = \frac{dK}{d\alpha}.$$

Integrating over α and using $K(0) = 0$ yields (5.6). Also $K'(\alpha) \geq 0$ implies $K(\alpha) \geq K(0) = 0$.

2.5.19. Since K_1 and K_2 are both solutions of (5.6) we have

$$A_1^*K_1 + K_1\bar{A} + C^*C = 0,$$

$$\bar{A}^*K_2 + K_2A_2 + C^*C = 0.$$

Subtracting yields $A_1^*K + KA_2 = 0$ with $K = K_1 - K_2$. Multiplying this last equation on the left by $e^{A_1^*t}$ and on the right by $e^{A_2^*t}$ yields

$$K = -\int_0^\infty d(e^{tA_1^*}Ke^{tA_2}) = -\int_0^\infty e^{tA_1^*}(A_1^*K + KA_2)e^{tA_2}\, dt = 0.$$

2.5.21. $\bar{G}(s)$ stable implies that $|\bar{G}(i\omega)x|^2$ is a bounded function of ω vanishing as $|\omega| \to \infty$ for each x. Then

$$\max_{|x|=1} \max_{\omega \in \mathbb{R}} |\bar{G}(i\omega)x| = \frac{1}{2\varepsilon} < \infty$$

which implies $I - \varepsilon^2 \bar{G}(i\omega)^*\bar{G}(i\omega) > 0$.

Chapter 3

3.1.1. If (A, B, C) is real, complex conjugation leaves the algebraic Riccati equation invariant so $K = \bar{K}$ by uniqueness.

3.1.2. Fix ω and let $\mathscr{F}_\omega$ be the collection of all subsets A of $[0, \infty) \times \Omega$ such that $A(\omega) = \{t | (t, \omega) \in A\}$ is in $\mathscr{B}([0, \infty))$. Then $\mathscr{F}_\omega$ is a σ-algebra and contains $A_1 \times A_2$ for all A_1 in $\mathscr{B}([0, \infty))$ and A_2 arbitrary. By definition of product, the result follows.

3.1.3. Fix $T \geq 0$ and set

$$x_n(t, \omega) = \sum_{k=0}^{n-1} x\left(\frac{k+1}{n} T, \omega\right) 1_{kT \leq nt < (k+1)T} + x(T, \omega) 1_{t \geq T}.$$

Each term in the sum is a product of an $\mathscr{F}_T$-measurable random variable with a $\mathscr{B}([0, T])$-measurable function of time. Thus $x_n(\cdot)$ is $\mathscr{B}([0, T]) \times \mathscr{F}_T$ measurable. Since $x(\cdot)$ is right continuous, $x_n(t, \omega) \to x(t, \omega)$ for all $0 \leq t \leq T$. Thus $x(\cdot)$ is progressively measurable.

3.1.4. Let $G_n = \{x | \text{there is a } y \in C \text{ with } d(x, y) < 1/n\}$. Then G_n is open and $G_1 \supset G_2 \supset \cdots$ intersects to C. Let $A = \{\tau \leq t\}$ and $A_{ns} = \{x(s) \in G_n\}$. By right continuity it follows that

$$B_n = \bigcup_{0 \leq s \leq t} A_{ns} = \left(\bigcup_{\substack{0 \leq s \leq t \\ s \in \mathbb{Q}}} A_{ns}\right) \cup A_{nt}$$

for all $n \geq 1$. Thus if we show that A is the intersection of the sets B_n, the result will follow. If ω is in the intersection of the sets B_n, then for all $n \geq 1$ there exist $0 \leq s_n \leq t$ such that $x(s_n, \omega) \in G_n$. Thus $\overline{(x(s), 0 \leq s \leq t)}$ intersects G_n for all n and hence intersects C. Thus $\tau(\omega) \leq t$, and ω is in A. Conversely, suppose that ω is in A but not in the intersection of the sets B_n. Then there is an N such that ω is not in B_N which implies for all $0 \leq s \leq t$, $x(s, \omega)$ is not in G_N. Since ω is in A, there exists $r_n \leq t + 1/n$ with $\overline{(x(s), 0 \leq s \leq r_n)} \cap C \neq \varnothing$. Thus there exists $0 \leq s_n \leq r_n \leq t + 1/n$ with $x(s_n, \omega)$ in G_{n+N}. Because ω is not in B_N, we must have $t < s_n$ for all n. Letting $n \uparrow \infty$, right continuity implies that $x(t, \omega)$ is in C, contradicting the fact that ω is not in B_N.

3.1.5. If $x(\cdot, \omega)$ is continuous, then

$$\overline{(x(s), 0 \leq s \leq t)} = (x(s), 0 \leq s \leq t).$$

This is enough to show that $\tau_C(\omega) = \tau(\omega)$.

3.1.6. We are given $x_n \to x$ in probability with $|x_n(\omega)| \leq c$. Then

$$P(|x| \geq 2c) \leq P(|x - x_n| \geq c) \to 0 \qquad \text{as} \quad n \uparrow \infty.$$

Thus x is a.s. bounded by $2c$. Now

$$E^P(|x - x_n|^r) = E^P(|x - x_n|^r; |x - x_n| < \varepsilon) + E^P(|x - x_n|^r; |x - x_n| \geq \varepsilon)$$

$$\leq \varepsilon^r + (3c)^r P(|x - x_n| \geq \varepsilon).$$

Letting $n \uparrow \infty$ and then $\varepsilon \downarrow 0$, the result follows.

3.1.7. If $Q(A_n) \nrightarrow 0$ then by passing to a subsequence we can assume that $Q(A_n) \geq \delta > 0$. By passing to another subsequence we can assume in addition that $P(A_n) \leq 2^{-n}$. Thus

$$P\left(\bigcap_{n=1}^{\infty} \bigcup_{m=n}^{\infty} A_m\right) = \inf_{n \geq 1} \sum_{m=n}^{\infty} P(A_m) \leq \inf_{n \geq 1} \sum_{m=n}^{\infty} 2^{-m} = 0$$

and so by absolute continuity

$$0 = Q\left(\bigcap_{n=1}^{\infty} \bigcup_{m=n}^{\infty} A_m\right) = \inf_{n \geq 1} Q\left(\bigcup_{m=n}^{\infty} A_m\right) \geq \inf_{n \geq 1} Q(A_n) \geq \delta.$$

Thus $Q(A_n) \to 0$. The second statement follows by taking $A_n = \{|x - x_n| \geq \varepsilon\}$.

3.1.8. Since x_n converges in L^r, x_n converges in P-probability; by 3.1.7, x_n converges in Q-probability. By 3.1.6, x_n converges in $L^r(\Omega, \mathscr{F}, Q)$.

3.1.9. For B fixed let $\mathscr{M}_B$ be the collection of sets A satisfying

$$P(A \cap B) = P(A)P(B).$$

Then, because both sides are measures in A, $\mathscr{M}_B$ is a σ-algebra. Since the collection of sets $\mathscr{M}$ independent of $\mathscr{F}_0$ is the intersection of the collections $\mathscr{M}_B$ over B in $\mathscr{F}_0$, $\mathscr{M}$ is a σ-algebra. It remains to be shown that $\eta(t)$ is $\mathscr{M}$ measurable for each $t \geq 0$. But this follows from

$$P(\eta(t) \in A | \mathscr{F}_0) = \int_A g_m(t - 0, x - \eta(0))\, dx = P(\eta(t) \in A) \quad \text{a.s. } P$$

and Exercise 3.1.11.

3.1.10. Let $\varphi(x) = \exp(cx)$ and let $\varphi_n \geq 0$ be a sequence of simple functions with $\varphi_n \uparrow \varphi$. Then by (iii) for B in $\mathscr{F}_s$ and $t > s$

$$E^P(e^{c\eta(t)}; B) = \lim_{n \uparrow \infty} E^P(\varphi_n(\eta(t)); B)$$

$$= \lim_{n \uparrow \infty} E^P\left(\int_{\mathbb{R}^m} g_m(t - s, x - \eta(s))\varphi_n(x)\, dx; B\right)$$

$$= E^P\left(\int_{\mathbb{R}^m} g_m(t - s, x - \eta(s))e^{cx}\, dx; B\right)$$

$$= E^P\left(\exp\left(c\eta(s) + \frac{|c|^2(t - s)}{2}\right); B\right).$$

Thus $R(\cdot)$ is a martingale. Conversely, suppose that $R(\cdot) = R_c(\cdot)$ is a martingale for all c. Then.

(*) $$E^P(R_c(t); B) = E^P(R_c(s); B)$$

for all $t > s \geq 0$ and B in $\mathscr{F}_s$. We need to show that (*) holds for complex vectors c. To this end note that

$$e^{|\eta(t)|} \leq \max\{e^{c\eta(t)} | c_j = \pm 1, j = 1, \ldots, m\}$$

is integrable and so $e^{c\eta(t)}$ is integrable uniformly in c in compact sets. Thus both sides of $(*)$ are continuous functions of the complex vector c. Since the integral of either side over a closed contour is zero, both sides are entire functions of c. Thus $(*)$ holds with c replaced by ic, $i = \sqrt{-1}$. By taking linear combinations

$$E^P(\varphi(\eta(t)); B) = E^P\left(\int_{\mathbb{R}^m} g_m(t - s, x - \eta(s))\varphi(x)\, dx; B\right)$$

for all trigonometric polynomials φ. Since both sides are measures, this implies that both sides are equal for all bounded measurable φ. Choosing $\varphi = 1_A$, the result follows.

3.1.11. $E^P(f|\mathscr{D}) = E^P(f)$ a.s. iff $E^P(f; B) = E^P(f)P(B)$ for all B in $\mathscr{D}$. Now f is independent of $\mathscr{D}$ iff there are simple functions f_n independent of $\mathscr{D}$ with $f_n \to f$. Since for f_n simple $E^P(f_n; B) = E^P(f_n)P(B)$ holds iff f_n is independent of $\mathscr{D}$, the result follows.

3.1.12. Assume first that g is bounded and choose g_n simple such that $g_n \to g$ boundedly. Then for B in $\mathscr{D}$,

$$\begin{aligned}
E^P(fg; B) &= \lim_{n\uparrow\infty} E^P(fg_n; B) \\
&= \lim_{n\uparrow\infty} E^P(E^P(f|\mathscr{D})g_n; B) \\
&= E^P(E^P(f|\mathscr{D})g; B).
\end{aligned}$$

The general case follows by noting that a.s. on $|g| \le n$

$$\begin{aligned}
E^P(fg|\mathscr{D}) &= 1_{|g|\le n} E^P(fg|\mathscr{D}) \\
&= E^P(1_{|g|\le n} fg|\mathscr{D}) \\
&= E^P(f|\mathscr{D})1_{|g|\le n} g \\
&= E^P(f|\mathscr{D})g.
\end{aligned}$$

3.1.13. By the mean value theorem

$$R_{\varepsilon c}(t) - 1 = R_{a\varepsilon c}(t)(c\eta(t) - a\varepsilon|c|^2 t)\varepsilon$$

for some $0 < a < 1$. Since $e^{k|\eta(t)|}$ is integrable for all $k > 0$ (see the solution of 3.1.10) and $d < e^d$ it follows that

$$\sup_{|c|+|d|\le r} c\eta(t)e^{d\eta(t)}$$

is integrable for all $r > 0$. Thus when differentiating $R_{\varepsilon c}(t)$ with respect to ε, we can take the limit inside the expectation. Thus for B in $\mathscr{F}_s$

$$\begin{aligned}
E^P(c\eta(t); B) &= E^P\left(\frac{d}{d\varepsilon}\bigg|_{\varepsilon=0} R_{\varepsilon c}(t); B\right) \\
&= \frac{d}{d\varepsilon}\bigg|_{\varepsilon=0} E^P(R_{\varepsilon c}(t); B) \\
&= \frac{d}{d\varepsilon}\bigg|_{\varepsilon=0} E^P(R_{\varepsilon c}(s); B) \\
&= E^P\left(\frac{d}{d\varepsilon}\bigg|_{\varepsilon=0} R_{\varepsilon c}(s); B\right) = E^P(c\eta(s); B).
\end{aligned}$$

Thus $c\eta(\cdot)$ is a martingale. Similarly, by differentiating twice we have $|c\eta(t)|^2 - |c|^2 t$, $t \geq 0$, is a martingale. Thus the second statement holds with $Q = c^*c$. Since any $Q = Q^*$ is a linear combination of matrices of the form c^*c, the result follows.

3.1.14. Let $x_n(\cdot) = x_T(\cdot)$ with $T = n$. Then the hypothesis of the Completeness Lemma holds trivially. Thus there is a unique progressively measurable right continuous process $x(\cdot)$ satisfying

$$\lim_{n\uparrow\infty} P\left(\sup_{0\leq t\leq T} |x_n(t) - x(t)| \geq \varepsilon\right) = 0$$

for all $T > 0$ and all $\varepsilon > 0$. Since $x_n(t) = x_T(t)$ for $0 \leq t \leq T$ and $n \geq T$, almost surely, the result follows.

3.1.15. For all $T > 0$, $\varepsilon > 0$, and $m \geq n$,

$$P\left(\sup_{0\leq t\leq T} |x_n(t) - x_m(t)| \geq \varepsilon\right)$$

$$\leq P\left(\sup_{0\leq t\leq T} |x_n(t) - x_m(t)| \geq \varepsilon \text{ and } T < \tau_n\right) + P(\tau_n \leq T)$$

$$= 0 + P(\tau_n \leq T).$$

Since $\tau_n \uparrow \infty$ as $n \uparrow \infty$, the hypothesis of the Completeness Lemma is satisfied. Thus there exists a progressively measurable right continuous process $x(\cdot)$ such that

$$P\left(\sup_{0\leq t\leq T} |x(t) - x_n(t)| \geq \varepsilon\right) \to 0 \qquad \text{as} \quad n\uparrow\infty.$$

By passing to a subsequence (see 3.2.1) $\{n_k\}$ we have

$$1 = P\left(\lim_{k\uparrow\infty} \sup_{0\leq t\leq T} |x_{n_k}(t) - x(t)| = 0\right)$$

$$\leq P\left(\lim_{k\uparrow\infty} \sup_{0\leq t<T\wedge\tau_n} |x_{n_k}(t) - x(t)| = 0\right)$$

$$= P\left(\lim_{k\uparrow\infty} \sup_{0\leq t<T\wedge\tau_n} |x_n(t) - x(t)| = 0\right)$$

$$= P\left(\sup_{0\leq t<T\wedge\tau_n} |x_n(t) - x(t)| = 0\right).$$

Letting $T \uparrow \infty$, the result follows.

3.1.16. We need to check (iii): for $t > s$ and B in $\mathcal{B}(\mathbb{R}^m)$

$$P(\eta(t) \in B | \mathcal{D}_s) = E^P(P(\eta(t) \in B | \mathcal{F}_s) | \mathcal{D}_s)$$

$$= E^P\left(\int_B g_m(t-s, x - \eta(s)) \, dx | \mathcal{D}_s\right)$$

$$= \int_B g_m(t-s, x - \eta(s)) \, dx \quad \text{a.s. } P,$$

since $\eta(s)$ is $\mathcal{D}_s$-measurable.

3.1.18. Define $x_n = |x(nT/N)|$, $n = 1, \ldots, N$. Then with $\mathscr{F}_n = \mathscr{F}_{nT/N}$

$$E^P(x_n | \mathscr{F}_m) = E^P\left(\left|x\left(\frac{nT}{N}\right)\right| \middle| \mathscr{F}_{mT/N}\right)$$

$$\geq \left|E^P\left(x\left(\frac{nT}{N}\right) \middle| \mathscr{F}_{mT/N}\right)\right| = \left|x\left(\frac{mT}{N}\right)\right| = x_m$$

whenever $n > m$. Applying 3.1.17 with $N = 2^n$, we have $x_N = |x(T)|$ and

$$P\left(\max_{1 \leq k \leq N} x_k \geq \lambda\right) \leq \frac{1}{\lambda} E^P\left(|x(T)|; \max_{1 \leq k \leq N} x_k \geq \lambda\right).$$

Now check that right continuity implies that $\max\{|x(kT/N)|: 1 \leq k \leq 2^n\}$ increases to $\sup\{|x(t)|: 0 \leq t \leq T\}$ as $n \uparrow \infty$.

3.1.19.

$$E^P(f^r) = E^P\left(\int_0^f r\lambda^{r-1}\, d\lambda\right)$$

$$= E^P\left(\int_0^\infty r\lambda^{r-1} 1_{f \geq \lambda}\, d\lambda\right) = r\int_0^\infty \lambda^{r-1} E^P(1_{f \geq \lambda})\, d\lambda = r\int_0^\infty \lambda^{r-1} P(f \geq \lambda)\, d\lambda,$$

$$E^P(f^r) = \int_0^\infty r\lambda^{r-1} P(f \geq \lambda)\, d\lambda \leq \int_0^\infty r\lambda^{r-1} \frac{1}{\lambda} E^P(g; f \geq \lambda)\, d\lambda$$

$$= \frac{r}{r-1} E^P\left(\int_0^f (r-1)\lambda^{r-2}\, d\lambda\, g\right) = \frac{r}{r-1} E^P(f^{r-1}g).$$

Now use Holder's inequality (Section 4.2) with the roles of f and g the reverse of what appears in Section 4.2. This yields

$$E^P(f^r) \leq \frac{r}{r-1} (E^P(g^r))^{1/r} (E^P((f^{r-1})^{r'}))^{1/r'}.$$

Since $(r-1)r' = r$, dividing out by $E^P(f^r)^{1/r'}$, the result follows.

3.1.20. Combine 3.1.18 with 3.1.19.

3.1.21. Since $x(\cdot)$ is a martingale,

$$0 \leq E^P(|x(t) - x(s)|^2) = E^P(|x(t)|^2) - 2E^P(x(t)^*x(s)) + E^P(|x(s)|^2)$$

$$= E^P(|x(t)|^2) - 2E^P(E^P(x(t)|\mathscr{F}_s)^*x(s)) + E^P(|x(s)|^2)$$

$$= E^P(|x(t)|^2) - E^P(|x(s)|^2).$$

Thus $E^P(|x(t)|^2)$ increases to a finite limit and (1.6) follows.

3.1.22. Let $\bar{\mathscr{F}}_t = \mathscr{F}_{t+s}$, $\bar{x}(t) = x(t+s) - x(s)$, $t \geq 0$. Then, by 3.1.18,

$$P\left(\sup_{s \leq t \leq T} |x(t) - x(s)| \geq \varepsilon\right) \leq \frac{1}{\varepsilon} E^P(|x(T) - x(s)|)$$

$$\leq \frac{1}{\varepsilon} (E^P(|x(T) - x(s)|^2))^{1/2}.$$

Letting $T \uparrow \infty$ and then $s \uparrow \infty$, (1.7) follows.

3.1.23. Let $\mathcal{M}_B$ be the collection of all sets in $\mathcal{F}$ independent of B, i.e., satisfying $P(A \cap B) = P(A)P(B)$. Because both sides of this last equation are measures, the collection of sets on which they agree forms a σ-algebra. Since $\mathcal{M}$ is the intersection of $\mathcal{M}_B$ over all B in $\mathcal{N}$, $\mathcal{M}$ is also a σ-algebra.

3.1.24. We have to check (iii), i.e., for all B in $\overline{\mathcal{F}}_s$ and $t > s$,

$$(*) \qquad P(\{\eta(t) \in A\} \cap B) = E^P\left(\int_A g_m(t - s, x - \eta(s)) \, dx; \, B \right).$$

We know that for B in $\mathcal{F}_s$ $(*)$ holds. Also if B is in $\mathcal{M}$, Exercise 3.1.11 shows that $(*)$ holds. Since the collection of sets B for which $(*)$ holds is a σ-algebra, $(*)$ holds for $\overline{\mathcal{F}}_s$.

3.1.25. For all $x \geq 0$, $\log^+(x) \leq (1/e)x$ (draw a picture). If $a \leq b$, the inequality holds. So assume $a > b$. If $1 \geq b \geq 0$, then the inequality follows by choosing $x = a$. So assume $a > b > 1$. In this case, the inequality follows by choosing $x = a/b$.

3.1.26. By 3.1.19 and 3.1.25, with $a = f$ and $b = g$

$$E^P(f) = \int_0^\infty P(f \geq \lambda) \, d\lambda \leq 1 + \int_1^\infty P(f \geq \lambda) \, d\lambda \leq 1 + E^P\left(\int_1^\infty g \frac{1}{\lambda} 1_{f \geq \lambda} \, d\lambda \right)$$

$$= 1 + E^P(g \log^+ f) \leq 1 + E^P\left(\frac{f}{e}\right) + E^P(g \log^+ g).$$

Transposing the second term to the left-hand side, the result follows.

3.1.27. Follows from 3.1.18 and 3.1.26.

3.1.28. Since $x(\cdot)$ is right continuous, so is $x_\tau(\cdot)$. Since $x_\tau(\cdot)$ is right continuous, all we need check is that $x(t \wedge \tau)$ is $\mathcal{F}_t$-measurable (3.1.3). Since $x(\cdot)$ restricted to $[0, t] \times \Omega$ is $\mathcal{B}([0, t]) \times \mathcal{F}_t$ measurable, all we need check is that the map $\omega \to (t \wedge \tau(\omega), \omega)$ is measurable $\mathcal{F}_t \to \mathcal{B}([0, t]) \times \mathcal{F}_t$. Since the second component of this map is the identity, this reduces to verifying the measurability of the first component. But $\{t \wedge \tau \leq a\}$ equals Ω if $t \leq a$ and equals $\{\tau \leq a\}$ if $a < t$. In either case this is $\mathcal{F}_t$-measurable. The result follows. We note that this result holds without right continuity. Moreover, τ need not be a stopping time, but need only satisfy $\{\tau < t\} \in \mathcal{F}_t$ for all $t \geq 0$. (Exercise 1.5.12 of [3.8].)

3.2.1. Set $A_n^k = \{\sup_{0 \leq t \leq T} |x_n(t) - x(t)| \geq 2^{-k}\}$. For each $k \geq 1$ choose n_k such that $P(A_{n_k}^k) \leq 2^{-k}$. By (2.9), this is possible. Then for each $m \geq 1$

$$P\left(\limsup_{j \uparrow \infty} \sup_{k \geq j} \sup_{0 \leq t \leq T} |x_{n_k}(t) - x(t)| \geq 2^{-m} \right)$$

$$\leq P\left(\bigcap_{j \geq m} \bigcup_{k \geq j} A_{n_k}^k \right)$$

$$\leq \inf_{j \geq m} \sum_{k \geq j} P(A_{n_k}) \leq \inf_{j \geq m} \sum_{k \geq j} 2^{-k} = \inf_{j \geq m} 2^{-j+1} = 0.$$

The result follows by taking the union over $m \geq 1$.

3.2.2. Set $|x(\cdot)|_2 = (E^P(\sup_{0 \leq t \leq T} |x(t)|^2))^{1/2}$. Then with $x(\cdot)$ and $x_j(\cdot)$ as in Section 3.2,

$$\left| x(\cdot) - \int_0^\cdot e(s) d\eta(s) \right|_2 \leq \lim_{j \uparrow \infty} |x(\cdot) - x_j(\cdot)|_2 + \left| x_j(\cdot) - \int_0^\cdot e(s) \, d\eta(s) \right|_2$$

$$\leq \lim_{j\uparrow\infty} 2E^P \left(\int_0^T |g_j(t) - e(t)|^2 \, dt \right)^{1/2}$$

$$= 2E^P \left(\int_0^T |g(t) - e(t)|^2 \, dt \right)^{1/2}.$$

Thus $x(\cdot)$ satisfies (2.6). If $x'(\cdot)$ also satisfies (2.6), the choosing $e(\cdot) = x_j'(\cdot)$ to approximate $x'(\cdot)$ as above, we see that $x(\cdot) = x'(\cdot)$.

3.2.3. Equation (2.4) follows from

$$||x(\cdot)|_2 - |x_j(\cdot)|_2| \leq |x(\cdot) - x_j(\cdot)|_2,$$

the fact that (2.4) holds for $x_j(\cdot)$, and (2.10). Since $x_j(t)$, $x_j(s)$ tend to $x(t)$, $x(s)$ in L^2 and $x_j(\cdot)$ is a martingale, it follows that $x(\cdot)$ is a martingale.

3.2.4. One way to show this is to use simple approximations. We outline another method following Exercise 4.6.9 of [3.8]. We use Doob's optional stopping theorem: if $x(\cdot)$ is an $(\Omega, \mathscr{F}_t, P)$ martingale then so is $x(\cdot \wedge t)$ and

$$E^P(x(t)|\mathscr{F}_{t\wedge\tau}) = x(t \wedge \tau) \quad \text{a.s. } P$$

for all $t \geq 0$. (For these facts as well as the definition of $\mathscr{F}_{t\wedge\tau}$ see Section 1.2 of [3.8].) Now, using 3.1.12 and 3.1.13, one can show that for simple $g(\cdot)$

$$\left| \int_0^t g(s) \, d\eta(s) \right|^2 - \int_0^t |g(s)|^2 \, ds, \qquad t \geq 0, \tag{1}$$

is a martingale. It follows then that (1) holds for all square integrable $g(\cdot)$. By Doob's theorem it follows that

$$\left| \int_0^{t\wedge\tau} g(s) \, d\eta(s) \right|^2 - \int_0^{t\wedge\tau} |g(s)|^2 \, ds, \qquad t \geq 0, \tag{2}$$

is also a martingale. Set $g_\tau(t) = 1_{t<\tau}g(t)$, $t \geq 0$; replacing $g(\cdot)$ by $g(\cdot) - g_\tau(\cdot)$ in (2) yields

$$0 = E^P \left(\left| \int_0^{t\wedge\tau} g(s) \, d\eta(s) - \int_0^{t\wedge\tau} g_\tau(s) \, d\eta(s) \right|^2 \right), \qquad t \geq 0. \tag{3}$$

Now setting $x(t) = \int_0^t g(s) \, d\eta(s)$, $t \geq 0$, we have

$$E^P(|x(t \wedge \tau) - x(t)|^2) = E^P(|x(t)|^2) - 2E^P(x(t)^*x(t \wedge \tau)) + E^P(|x(t \wedge \tau)|^2)$$

$$= E^P(|x(t)|^2) - E^P(|x(t \wedge \tau)|^2) - 2E^P((x(t) - x(t \wedge \tau))^*x(t \wedge \tau))$$

$$= E^P(|x(t)|^2) - E^P(|x(t \wedge \tau)|^2)$$

$$= E^P \left(\int_0^t |g(s)|^2 \, ds \right) - E^P \left(\int_0^{t\wedge\tau} |g(s)|^2 \, ds \right) \tag{4}$$

$$= E^P \left(\int_{t\wedge\tau}^t |g(s)|^2 \, ds \right).$$

Replacing $g(\cdot)$ by $g_\tau(\cdot)$ in (4) and combining this with (3), the result follows.

3.2.5. Set $f_n(t) = f(t)1(\int_0^t |f(s)| \, ds \leq n)$; this is progressively measurable and

$$\int_0^\infty |f_n(t, \omega)| \, dt \leq n \quad \text{for all } \omega.$$

Thus $x_n(t) = \int_0^t f_n(s)\, ds$, $t \geq 0$, is progressively measurable and right continuous. Clearly, the hypothesis of the Completeness Lemma holds. Thus there is a progressively measurable right continuous limit $x(\cdot)$. By passing to a subsequence as in 3.2.1, we can assume that $x_n(\cdot)$ converges to $x(\cdot)$ uniformly on $[0, T]$, almost surely. Thus

$$P\left(x(t) = \int_0^t f(s)\, ds, t \geq 0\right) = 1.$$

Actually one can find a version $x(\cdot)$ such that $x(t, \omega)$ equals $\int_0^t f(s, \omega)\, ds$ for all (t, ω) satisfying $\int_0^t |f(s, \omega)|\, ds < \infty$. To prove this, look at the proof of the Completeness Lemma.

3.2.6. Apply the Ito rule with $n = 1$, $f(\cdot) = -z(\cdot)^*$, $g(\cdot) = -|z(\cdot)|^2/2$, $x(\cdot) = \log R(\cdot)$, and $\varphi(x) = e^x$.

3.2.7. Let τ_n be the contact time of the right continuous version of $\int_0^{\cdot} |g_n(t)|^2\, dt$ with $\{x \mid |x| \geq 1\}$. Then

$$P(\tau_n \leq T) = P\left(\int_0^T |g_n(t)|^2\, dt \geq 1\right) \to 0 \qquad \text{as} \quad n \uparrow \infty.$$

Let $g_n'(t) = g(t)1_{t<\tau}$, $t \geq 0$, $\tau = \tau_n$; then $\int_0^\infty |g_n'(t)|^2\, dt \leq 1$ and so 3.1.6 implies that $\int_0^T |g_n'(t)|^2\, dt \to 0$ in the mean and hence, by 3.1.18, the result holds for $g_n'(\cdot)$. Since, by 3.2.4,

$$P\left(\sup_{0 \leq t \leq T} \left|\int_0^t g_n(s)\, d\eta(s)\right| \geq \varepsilon\right)$$

$$\leq P\left(\sup_{0 \leq t \leq T} \left|\int_0^t g_n'(s)\, d\eta(s)\right| \geq \varepsilon \text{ and } T < \tau_n\right) + P(\tau_n \leq T),$$

the result follows.

3.3.5. The proof of 3.3.3 is unchanged.

3.3.6. In going from $I^n(\cdot)$ to $J^n(\cdot)$: representing $I^n(\cdot)$ as

$$I^n(t, \omega) = F(t, y^n(t \wedge t_1, \omega), y^n(t \wedge t_2, \omega), \ldots), \qquad \omega \in \Omega, \quad t \geq 0,$$

set

$$J^n(t, \omega) = F(t, y(t \wedge t_1, \omega), y(t \wedge t_2, \omega), \ldots)1_{t<\sigma(\omega)}, \qquad \omega \in \Omega, \quad t \geq 0, \quad \sigma = \sigma_n.$$

Then $J^n(\cdot)$ is right continuous and $\mathscr{Y}_t$-progressively measurable.

Chapter 4

4.1.5. Direct computation.

4.1.9. Let $a(\cdot)$ denote the left-hand side and $b(\cdot)$ the right-hand side. Since $l_k(\cdot)$ satisfies

$$P\left(l_k(t) = 1 + \int_0^t l_k(s)z_k(s)^*\, dy(s), t \geq 0\right) = 1$$

multiplying by π_k^0 and summing over k yields

$$a(t) = \sum_{k=1}^{N} l_k(t)\pi_k^0 = 1 + \int_0^t \left(\sum_{k=1}^{N} \pi_k^0 l_k(s) z_k(s) \right)^* dy(s)$$

$$= 1 + \int_0^t a(s)\hat{z}(s)^* \, dy(s), \qquad t \geq 0, \quad \text{a.s. } P.$$

Moreover, the Ito rule yields

$$b(t) = 1 + \int_0^t b(s)\hat{z}(s)^* \, dy(s), \qquad t \geq 0, \quad \text{a.s. } P.$$

Applying the Ito rule to $\varphi(a, b) = a/b$, the result follows.

4.2.2. If (a) holds then with $\varphi_j(k) = \delta_{jk}$, $P(\pi_j(\infty) = \delta_{j\theta}) = 1$. Hence

$$P(\pi_\theta(\infty) = 1) = \sum_{j=1}^{N} P(\pi_\theta(\infty) = 1, \theta = j)$$

$$= \sum_{j=1}^{N} P(\pi_j(\infty) = 1, \theta = j)$$

$$= \sum_{j=1}^{N} P(\delta_{j\theta} = 1, \theta = j)$$

$$= \sum_{j=1}^{N} P(\theta = j) = 1,$$

so (b) holds. If (b) holds then clearly (a) holds. Since the left-hand side in (a) is $\mathscr{Y}_\infty$-measurable, (a) implies (c). Conversely, if (c) holds then there is a $\mathscr{Y}_\infty$-measurable random variable δ such that $\theta = \delta$ a.s. Then $\hat{\phi}(\infty) = E^P(\varphi(\theta)|\mathscr{Y}_\infty) = E^P(\varphi(\delta)|\mathscr{Y}_\infty) = \varphi(\delta) = \varphi(\theta)$ a.s.

4.2.7. Suppose first that $z(\cdot)$ satisfies (1.8) of Chapter 4; then 3.1.18 and 3.2.7 imply that

$$P\left(\sup_{0 \leq t \leq T} \exp\left(-\int_0^t z(s)^* \, d\eta(s) - \tfrac{1}{2} \int_0^t |z(s)|^2 \, ds \right) \geq \lambda \right) \leq \frac{1}{\lambda}.$$

Now let $\tau_n \uparrow \infty$ be stopping times such that $z^n(\cdot) = 1_{\cdot < \tau} z(\cdot)$, $\tau = \tau_n$, satisfies (1.8) with $c = n$; then

$$P\left(\sup_{0 \leq t \leq T \wedge \tau} \exp\left(-\int_0^t z(s)^* \, d\eta(s) - \tfrac{1}{2} \int_0^t |z(s)|^2 \, ds \right) \geq \lambda \right) \leq \frac{1}{\lambda}.$$

The result follows upon letting $n \uparrow \infty$, $T \uparrow \infty$, and $\lambda \uparrow \infty$, in that order.

4.2.8. In the notation of 4.2.6, each $L_k(\cdot)$ is an exponential martingale and so 4.2.7 applies. Since $\pi_\theta(\cdot)^{-1}$ is a linear combination of the $L_k(\cdot)$'s, it follows that

$$P\left(\sup_{t \geq 0} \pi_\theta(t)^{-1} < \infty \right) = 1.$$

4.2.9. Proceeding as in 4.2.1, we see that θ is $\mathscr{Y}_T$-measurable iff $P(\pi_\theta(T) = 1) = 1$. Since $l_j(T) > 0$ for all j, this happens iff $P(\pi_\theta^0 = 1) = 1$. This implies $\pi_j^0 = P(\theta = j) = E^P(\pi_\theta^0, \theta = j) = P(\pi_j^0, \theta = j) = \pi_j^{0^2}$ for all j which implies that θ is a.s. constant.

4.2.10. Since $y(\cdot)$ is an $(\Omega, \mathscr{F}_t, Q)$ Brownian motion and θ is $\mathscr{F}_0$-measurable, θ and $y(\cdot)$ are Q-independent. Thus θ $\mathscr{Y}_\infty$-measurable implies θ is Q-independent of itself and hence θ is Q-almost surely constant. Since $Q = P$ on $\mathscr{F}_0$, the result follows.

4.2.11. Consistency implies that θ is $\mathscr{Y}_\infty^P$-measurable. Since $Q \ll P$, θ is $\mathscr{Y}_\infty^Q$-measurable. Thus θ is Q-independent of itself and the result follows.

4.3.1. Since $y(t, \omega) = (y_T(\omega))(t)$ for $0 \le t \le T$, $y(t)$ is $y_T^{-1}(\mathscr{M}_T)$-measurable for all $0 \le t \le T$ and hence $\mathscr{Y}_T \subset y_T^{-1}(\mathscr{M}_T)$. Conversely, if f is $\mathscr{M}_T$-measurable then

$$f(\alpha) = F(\alpha(t_1), \alpha(t_2), \ldots)$$

for some $0 \le t_j \le T$. Thus $f(y_T(\omega)) = F(y(t_1, \omega), y(t_2, \omega), \ldots)$ is $\mathscr{Y}_T$-measurable. The case $T = \infty$ is similar.

4.3.3. Check that

(*)
$$\sum_{k=1}^N \pi_k \log\left(\frac{\pi_k}{\pi_k^0}\right) = \sup\left\{\sum_{k=1}^N v_k \pi_k - \log\left(\sum_{k=1}^N e^{v_k} \pi_k^0\right)\right\},$$

where the supremum is over all real v_j. Choosing $v_j = 0$, we see that

$$0 \le \sum_{k=1}^N \pi_k \log\left(\frac{\pi_k}{\pi_k^0}\right) = \sum_{k=1}^N \pi_k \log \pi_k - \sum_{k=1}^N \pi_k \log \pi_k^0.$$

Choosing $\pi_k^0 = 1/N$, the result follows. Alternatively we know from Section 4.3 that $I(\pi; \pi^0) \ge 0$. However, we did it this way to introduce (*), which is frequently used as the definition of $I(\pi, \pi^0)$, as the right-hand side makes sense whether or not we have absolute continuity.

4.3.6. By 4.3.4, $I(\infty)$ is as large as possible iff

$$P\left(\sum_{k=1}^N \pi_k(\infty)\log \pi_k(\infty) = 0\right) = 1$$

which happens iff $\pi_j(\infty) = \delta_{jk}$ for some random k. But 4.2.8 says that $\pi_\theta(\infty) > 0$. Thus $k = \theta$ and the result follows.

Chapter 5

5.2.2. Let

$$\delta(t) = \pi(t) \exp\left(-\int_0^t f(s)\,ds - \int_0^t z(s)^*\,d\eta\,(s) + \tfrac{1}{2}\int_0^t |z(s)|^2\,ds\right), \qquad t \ge 0.$$

Using the Ito rule check that $P(\delta(t) = \delta(0), t \ge 0) = 1$. Since here $\delta(0) = 0$, the result follows.

5.2.3. Let $f'(t) = 1_{t < \zeta} f(t)$, $z'(t) = 1_{t < \zeta} z(t)$, and $\pi'(t) = \pi(t \wedge \zeta)$, $t \ge 0$. Then

$$P\left(\pi'(t) = \pi^0 + \int_0^t \pi'(s)f'(s)\,ds + \int_0^t \pi'(s)z'(s)^*\,d\eta\,(s), t \ge 0\right) = 1.$$

Defining $\delta'(\cdot)$ analogously to $\delta(\cdot)$, Ito's rule yields $P(\delta'(t) = \pi^0, t \ge 0) = 1$. The result follows.

5.2.4. Follows from the fact that a closed system of stochastic differential equations has a unique solution (see Section 3.2).

5.3.2. Follows from 2.3.7.

5.3.3. By (3.6), (3.7), and 1.3.8,

$$E^P\left(\int_0^\infty |u(t)|^2\,dt\right) \le cE^P\left(\int_0^\infty |\hat{z}(t)|^2\,dt\right) \le cE^P\left(\int_0^\infty |z(t)|^2\,dt\right)$$

$$\le c'E^P\left(\int_0^\infty |x(t)|^2\,dt\right) \le c''\left(1 + E^P\left(\int_0^\infty |v(t)|^2\,dt\right)\right) < \infty.$$

5.5.6. Direct computation.

Index

Absolutely continuous 46
Adaptive
— control 102
— stabilization 85
Admissible control 84, 94
Algebraic Riccati equation 28
Allinger 82
Almost surely continuous 45
Ash 63

Bar-Shalom 102
Bellman equation 99
Borel σ-algebra 43
Bounded real 37
—, Strictly 40
Brockett 20
Brownian motion 47
Byrnes 102

Cameron–Martin–Girsanov theorem
    58
Cayley–Hamilton theorem 6
Completeness lemma 46, 62
Concatenation 9
Conditional
— expectation 47
— probability 47
Consistency 73
Contact time 45
Control
—, Admissible 84, 94

—, Finite 26
—, Optimal 24, 95
— process 84
—, Stabilizing 21, 85
Controllable system 11
Convergence
— in L^r 46
— in probability 45
— in the mean 46
Cost 25, 95
—, Optimal 25

Davis 102
Delchamps 41
Denominator 6
Doob 49, 50, 63
Duncan 82
Dunn 82
Durrett 82

Eigenpolynomial 4
Entrance time 45
Equivalent systems 9
Event space 43
Expectation 45
—, Conditional 47
Explosion time 57
Exponential
— martingale 56
— matrix 6

Feedback
— law　21, 86
—, Stabilizing　21
Fel'dbaum　102
Fleming　42
Fujisaki　83

Gauss–Weierstrass kernel　47
Ghosh　102
Gilbert　20

Hamiltonian　15
Hazewinkel　20
Hermann　20
Hirsch　20
Ho　20
Holder inequality　73

Ikeda　63
Impulse response　9
Independence　47
Inequality
—, Doob's　49, 50
—, Holder's　73
—, Jensen's　78
—, Young's　20
Information
— functional　78
— rate　82
—, Shannon　79
Innovations
— problem　72
— process　69
Input　8
Input–Output (I/O) map　9
Integrable　45
—, Uniformly　46
Ito　52, 63
— differential rule　55

Kallianpur　83
Kalman　20, 41
Krylov　42
Kumar　102
Kunita　83

Linear system　8
Lions　42
Liptser　63
Localizable　61
LQ regulator　21, 41
—, Adaptive　87
L^r　45
Luenberger　42
Lyapunov equation　8

Martin　20
Martingale　47
— convergence theorem　50
— inequality　49
— $L \log L$ inequality　51
— L^r inequality　50
Matrix exponential　6
McKean　63
Measurable　43
Minimal polynomial　6
Minimal system　13
Mitter　82
$M(m, n, p)$　35

Narendra　20
Nelson　63
Noise process　64
Nonanticipating　44

Observable
— eigenvalue　16
— system　12
Observations process　64
Optimal
— control　24, 95
— cost　25
Output　9

Pontrjagin　20
Probability
—, Conditional　47
— measure　45
— space　45
Process　43
—, Innovations　69
—, Noise　64
—, Observation　64
—, Signal　64
—, Simple　53
Product σ-algebra　43
Progressively measurable　44, 59
Proper　26

Random variable　45
Realization　9
Resolvent　5
Riccati equation
—, Algebraic　28
—, Differential　33
Right continuous　44
Rishel　42
Rudin　20, 42

σ-algebra
—, Borel   43
— generated by a process   44
—, Product   43
Shannon information   79
Shayman   42
Shiryayev   63
Smale   20
Stable
— matrix   7
— polynomial   7
— transfer function   15
Stabilizable   22
—, Adaptively   92
—, Uniformly   92
Stabilizing
— control 21, 85
— feedback   22
Standard controllable form   16
State trajectory   8
Stein   20
Stochastic
— calculus   52
— differential equations   57
— integrals   52

Stopping time   44
Streibel   83
Stroock   63

Tannenbaum   42
Transfer function   9
Triple   8

Uniformly
— bounded   46
— integrable   46
— stabilizable   92

Varadhan   63

Watanabe   63
Weiss   20
Wiener   63
— measure   48
— space   48
Willems   83

Young's inequality   20
$\mathcal{Y}_t$-progressively measurable   59